Praise for *Plants*...

'A passionate invitation to [...] botanical world. With warm [...] Fletcher and Head blend cult[...] ...wledge, eliciting rich narratives from u[...], sediment cores and old photographs. A book about science, history, invasion and the Anthropocene, *Plants: Past, Present and Future* calls for new ways of understanding and engaging with Country and reveals the power and possibility of Indigenous ecological expertise.'
— Billy Griffiths

'From grasslands to yams and fruit-bearing trees, this book highlights the importance of Indigenous knowledge. An enlightening read on the power of plants and the management practices of Indigenous people.'
— Terri Janke

Danielle Gorogo, *Treelines*, 2008

Treelines, the artwork detail used on the cover and reproduced in full above, shows an aerial view of Country, of the trees lining the banks of rivers meandering through the landscape. The relationship between Indigenous people and trees is one of coexistence. A living tree was usually marked for spiritual and historical significance; Treelines, like Songlines, are maps of the land Aboriginal people live on. People sing as they pass through Country, sing the stories of that Country and their relationship to it, and Treelines form part of this story.

Danielle Gorogo is a Clarence Valley First Nations artist living in the Northern Rivers region of New South Wales. She is a direct descendant of the Dunghutti, Gumbaynggirr and Bundjalung nations. Danielle's multifaceted cultural heritage, which includes First Nations Australian, Papua New Guinean, Māori and Micronesian ancestry, is reflected in her art.

PLANTS

Aboriginal and Torres Strait Islander peoples are advised that this book contains the names and images of people who have passed away.

The stories in this book are shared with the permission of the original storytellers.

PLANTS

Past, Present and Future

ZENA CUMPSTON,
MICHAEL-SHAWN FLETCHER
& LESLEY HEAD

Thames &Hudson | national museum australia

First published in Australia in 2022
by Thames & Hudson Australia Pty Ltd
11 Central Boulevard, Portside Business Park
Port Melbourne, Victoria 3207
ABN: 72 004 751 964

thamesandhudson.com.au

Plants © Thames & Hudson Australia 2022

Introduction © Margo Neale/NMA 2022
Text © Zena Cumpston, Michael-Shawn Fletcher and Lesley Head 2022
Images © copyright remains with the individual copyright holders

25 24 23 22   5 4 3 2

The moral right of the authors has been asserted.

All rights reserved. No part of this publication may be reproduced or transmitted in any form or by any means, electronic or mechanical, including photocopy, recording or any other information storage or retrieval system, without prior permission in writing from the publisher.

Any copy of this book issued by the publisher is sold subject to the condition that it shall not by way of trade or otherwise be lent, resold, hired out or otherwise circulated without the publisher's prior consent in any form or binding or cover other than that in which it is published and without a similar condition including these words being imposed on a subsequent purchaser.

Thames & Hudson Australia wishes to acknowledge that Aboriginal and Torres Strait Islander people are the first storytellers of this nation and the traditional custodians of the land on which we live and work. We acknowledge their continuing culture and pay respect to Elders past, present and future.

ISBN 978-1-760-76187-5 (paperback)
ISBN 978-1-760-76188-2 (ebook)

A catalogue record for this book is available from the National Library of Australia

Every effort has been made to trace accurate ownership of copyrighted text and visual materials used in this book. Errors or omissions will be corrected in subsequent editions, provided notification is sent to the publisher.

This project has been assisted by the Australian Government through the Australia Council, its arts funding and advisory body.

Front cover: *Treelines* by Danielle Gorogo

Series editor: Margo Neale
Cover design: Nada Backovic
Typesetting: Megan Ellis
Editing: Declan Fry

Printed and bound in Australia by McPherson's Printing Group

FSC® is dedicated to the promotion of responsible forest management worldwide. This book is made of material from FSC®-certified forests and other controlled sources.

*For my mum and dad, Noelene Zada and Trevor Cumpston.
Journeying with me, always.*
ZC

*For my family, whose unwavering support and love light my path,
making this all possible.*
MSF

To Biddy Simon, and to the memory of the late Polly Wandanga.
LH

NOTE ON STYLE AND SPELLING

Readers may note that for different language groups, or within language groups, variant spellings occur for similar words, cultural groups or names. Some of the variations encountered this book include:

Barka, Baaka

Barkandji, Bakandji, Barkindji, Paakantyi

Boon Wurrung, Bunurong

Murrinh-patha, Murrinhpatha

Njeri Njeri, Nyeri Nyeri

Woiwurrung, Woi Wurrung

CONTENTS

First Knowledges: An Introduction *Margo Neale*   1

1. Personal Perspectives *Zena Cumpston, Michael-Shawn Fletcher & Lesley Head*   9
2. Looking Back, Moving Forward *Zena Cumpston*   30
3. Bolin Bolin *Michael-Shawn Fletcher*   50
4. Abundance *Zena Cumpston*   69
5. Cumbungi *Lesley Head*   100
6. Yams *Lesley Head*   114
7. Spinifex *Lesley Head*   130
8. Quandongs *Zena Cumpston & Lesley Head*   149
9. Respecting Knowledge *Zena Cumpston*   160
10. Futures *Zena Cumpston & Michael-Shawn Fletcher*   173

Acknowledgements   179
Image Credits   181
Notes   183
Index   204

FIRST KNOWLEDGES

MARGO NEALE, SERIES EDITOR

Australian indigenous plants offer a portal through which we can learn of the deep knowledge and complex systems of land management practised by Aboriginal and Torres Strait Islander peoples over millennia. They offer us one of many ways of connecting to Country; of being one with it. Country is not only the heartbeat of this continent but also *our* heartbeat. It tells us who we are, how we should live, how to care for each other and care for Country. It holds the answer to our future survival on this planet.

Plants are not only part of Country – in our worldview they *are* Country. As such, plants are part of the Dreaming, given to us by the ancestors to ensure their survival. Plants have their own repertoire of song cycles practised by custodians in a variety of ways, as part of this cycle of survival. The knowledge imparted in these pages makes this book a tool for survival too, as the writers become part of the custodial circle revealing the many ways we can care for Country in the altered reality of the 21st century. You as reader, being complicit in this knowledge exchange, carry some responsibility for protecting Indigenous knowledges. The discussion of plants in this book is not just confined to food and medicine, nor is it confined to the most visible parts of plants – the foliage and flowers. It also concerns the wood used for making tools and weapons, the resin used to haft stone axe heads onto wooden shafts, building materials for shelters and fibres for making nets and carry bags, as well as the hulls of canoes and adornments for ceremony.

All of the books in this series are grounded in Country. It is the spine that runs through each of them and binds them as one, not unlike Songlines, which are storehouses of knowledge that run through time and place. In combination, this series of books on first knowledges takes the reader on a journey through an integrated knowledge system, a pathway along which they encounter multiple sites of learning. This bears comparison to how people can travel the Songlines, visualised as pathways or corridors that link sites of knowledge, in much the same way that these books do. Each explores the different ways that First Peoples connect with and experience Country. A relationship with plants is just one such way. Plants are also the doorway through which all peoples can come and join us on Country.

Each of the three authors of this publication brings original and enriching insights to the subject. Similarly, each brings assessments of the contemporary cultural landscape based on their own experiences navigating the multidimensional world of Aboriginality, whether through academia or lived experience. Their work here is a deliberate deviation from botany, though related in a limited way, as the writers untether plants from their disciplinary moorings to reveal the expansiveness of Indigenous knowledges concerning them. Conversely, other Western disciplinary titles used in the series, such as 'design', 'astronomy' and 'law', intentionally present a challenge. They demonstrate how knowledge is compartmentalised in the Western system, standing in strong contrast to the integrated systems of

Indigenous approaches, where there are no such separations. Views throughout this series diverge from one another, as they should; this series is as much about providing a forum for multiple voices, revealing a fluid and unsettled space, as it is about inviting readers to understand and appreciate Indigenous knowledges.

In keeping with this, the writers here include more than the visible and the objective. They dig below the surface, in multiple literal and metaphoric ways, to expose the myriad lives of plants and restore their cultural, ancestral and spiritual dimensions. In the process they widen the lens to reclaim urban areas as Country and reintegrate these Westernised spaces back into the deep foundations of Country, finding ways to 'reinscribe this connection into our everyday life'. As Melbourne-based researcher, educator and storyteller Zena Cumpston notes, the cultural context of Country is not 'lost' in urban areas. Though Country may lie dormant under certain circumstances, as it does beneath layers of concrete, bitumen and buildings for long periods of time, it always remembers. And once reactivated and acknowledged through appropriate human interaction, it is revitalised and resumes life. Out of sight is not out of mind. You may have heard the Aboriginal expression, 'Country in mind'. All places in Australia, whether urban or remote, are Country. As Zena notes, 'no matter how violent the alterations, Country still holds her stories and we each, no matter what our culture or race, have a role to play in keeping her healthy and strong'.

This book also reclaims our indigenous plants that have been deemed weeds by the colonisers, especially some of those growing in urban areas, struggling through the cracks in rocks and buildings, abandoned lots and suburban yards. Europeans turned *Typha*, or bulrushes, which we know as cumbungi, into weeds when our firing and management regimes were stopped and their distribution left uncontrolled. This allowed cumbungi to encroach into open waters, such as dams, wetlands and irrigation channels. Beyond its value as food, cumbungi is used in other parts of the world for insulation and wall construction. It is considered by author Lesley Head as a botanical 'canary in the coalmine', with its ability to both detect pollution and clean it up. Yet it has been labelled a weed to be eradicated.

The Western concept of Australia as wild and untamed wilderness is also overturned, the evidence showing it to be a mosaic of management constructed by Aboriginal people and not simply by climate. Fire was employed to manipulate plants and create new ecosystems, including grasslands, fruit tree groves and yam landscapes. The scale and nature of the creation of these fruitful cultural landscapes was, as Wiradjuri scientist Michael-Shawn Fletcher states, 'a firm and empirical rejection of the racist and dehumanising notion of "terra nullius"'. He sets about using the 'Western scientific method to prove our agency in creating and maintaining our Country'.

Michael-Shawn Fletcher specialises in palaeoecology and the long-term interaction between humans, climate and vegetation. To him, plants are 'the architects of life on Earth via their unique ability to convert the Sun's rays into food', and their capacity to exploit animals and humans despite being apparently unable to move. While the Western worldview traditionally considers humankind 'the masters of Earth', Michael proposes that, in reality, we may 'merely be vessels for the world domination of plants', pointing out that wheat 'now gets us to do its evolving for it'.

Lesley Head, a well-established ethnobotanist and physical and cultural geographer, draws on archaeology and various other fields of ecology for her contribution. Her overriding interest is in the span of peopled landscapes from tens of thousands of years ago to the suburban backyards of today, ensuring that the reader gains a full conceptual and cultural experience of that little word, 'plants'. She brings into view the overlooked issue of women's engagement in plant harvesting and preparation as a consequence of the emergence of feminist writers and thinkers who 'provided strong critiques of the "Man the Hunter" trope, stimulating researchers to consider how both women and plants ("Woman the Gatherer") had been erased from the long arc of human history'. That is, as written by Western historians. However, what women gathered and harvested is well recorded in the land, in the ancient archives preserved in rock art and oral histories. It did not help that the evidence of their work in the form of seeds and plant implements didn't preserve well, which gives weight to the expression that absence of evidence is not evidence of absence.

However, stop the presses. There is recent research on grindstones that shows evidence of breadmaking some 65,000 years ago in the Top End of Australia.[1] Evidence that the first women of Australia were the world's first breadmakers challenges the longstanding view that the cradle of civilisation was in the Middle East, where there is evidence of breadmaking (a sourdough-like bread made from wild grains) in Jordan some 14,000 years ago.

This is the fifth book in the First Knowledges series. It connects well with the previous book in the series, *Astronomy: Sky Country*. While *Plants* focuses on the terrestrial domain, at least in the literal sense, *Astronomy* takes us to the celestial, though both are inextricably related in our way of thinking. The first book in the series, *Songlines: The Power and Promise*, is as foundational to this series as Songlines are to Country and culture, revealing the basis of our knowledge system archived in the land. The second book, *Design: Building on Country*, explores how our objects and built structures are extensions of Country, alive and invested in varying degrees with the spirits of our ancestors. *Country: Future Fire, Future Farming*, the third book, addresses some critical contemporary issues with ideas about how we can manage Country sustainably and relearn how to live in respectful collaboration with the land, and not as usurpers.

Plants and *Country* are, naturally, closely related, as are all disciplines in our integrated knowledge system. Like the authors of *Country*, Bill Gammage and Bruce Pascoe, these writers call for action and demand that we look afresh at our continent and its vegetation through Australian eyes, and not continue the destructive practices of the colonisers. They implore us to bring back Indigenous land management systems and vegetate with indigenous plants adapted to this climate and soil, those that are not demanding of water, fertiliser and pesticides. They also urge collective action towards sustainable Indigenous food security and respectful partnerships, equitable benefit sharing and protections for the intellectual property rights of

First Australians, whose knowledges kept this continent healthy for some 60,000 years.

A defining characteristic of this series – and one that is reflected in the title of this book, *Plants: Past, Present and Future* – is the writers' regard for and application of traditional Indigenous practices and belief systems to contemporary circumstances for a renewed future. As Lesley notes, it is our hope that this book provides insights into the rich variety of Indigenous relations with, and knowledge about, plants, and ideas for how we can all live with plants symbiotically and sustainably on our increasingly fragile planet.

Michael-Shawn Fletcher makes a revelatory observation: 'If a keen eye, a keen mind and a five-year journey around Earth on the *Beagle* had inspired one of the most, if not *the* most, important (Western) scientific advances – Charles Darwin's theory of evolution – how much science must reside in the knowledge bank of Aboriginal Australia? What revolutions lie within our knowledge systems?' It is in a book series such as this, and innumerable other projects, exhibitions and programs on Country and in the bush tucker realm, as well as language revival, that this wisdom continues to surface year after year. There is nowhere else on Earth that can claim the long human occupancy that we can claim on this continent. The rest of the world will catch up with this fact over time as the globe threatens to reel out of control and desperate bids are made to reconnect with our ancient pasts for guidance. It is now not uncommon for people to look to Indigenous knowledges, particularly in the area of climate change and environment. As the expression goes, 'When you look behind, you see the future in your footsteps.'

1

PERSONAL PERSPECTIVES

ZENA CUMPSTON

I am an Aboriginal woman with Afghan, Irish and English heritage. My people are the Barkandji from western New South Wales. My mum brought me up strong, to be proud to be an Aboriginal person.

I have fond memories of our regular trips to the bush to visit Mum's big family – Broken Hill, Menindee, Winnathee, Cobar. Our 1975 Ford Falcon station wagon would too often make ominous noises and Mum would crank the radio up to cover the annoyance. Somehow we always made it. I don't remember the green garbage machine ever breaking down or running out of petrol. That car made it a long way despite not having much in the tank, technology exquisitely matched to the modus of our beautiful mum.

For Barkandji mob, every part of our lives is tied to our Barka (also sometimes Baaka) – known to many as the 'Darling River'. This deep connection is reflected in our name, 'Barkandji' meaning 'people belonging to the river'. I am a Barkandji wiimpatja, or Darling River Black person. Lots of mob spell our name lots of different ways – there is no one 'right way' because our languages were not written – but if you want to begin to know about Barkandji Country or any Barkandji person, then you need to know about our Barka. The statement given by Uncle Badger Bates, a Barkandji elder, artist, activist and knowledge holder, to the Murray-Darling Basin Royal Commission, held in South Australia in 2018, is a good place to start.[1] Terrible things have happened to it in recent times – water theft, a lack of respect, and sheer greed and mismanagement have put our lifeblood at great risk. We could lose it forever. The commodification of water and lack of water rights for Indigenous communities isn't just a problem for my mob, it can and is happening around the country.

I have always been interested in plants, but my real passion began when I was early in my pregnancy with my first child, Lou. I got a strong message to learn to garden. This directive is almost impossible to explain in words. It came from somewhere I can't place, from inside me, but it felt as if it was put there by someone or something else. It was forceful and cerebral but it sat in my gut more than my head. It was an absolute knowing more than a suggestion. Make a garden, it said, wordlessly. And so, with help from my friends, and despite living in a unit surrounded by concrete, I built a beautiful little garden. Over time (and through my gardening) I became more and more interested in the ways my people have interacted with plants. I started reading and thinking and yarning and reading some more. I began to develop a powerful sense of the abundance that is possible through plants, to understand how much of our capacity to thrive and to innovate came from our ability to interact with plants, to know them intimately. And then things began to fall into place, as they do when your ancestors are guiding you, and more than once I managed to somehow be in the right place at exactly the right time.

Despite not having a PhD, and testament to some of the great new opportunities some universities are providing to make room for more mob, I was successful in attaining a research fellow position for the Clean Air Urban Landscapes (CAUL) Hub. The CAUL Hub (2015–21) was a consortium funded by the Australian Government's National Environmental Science Program (NESP). It included researchers from multiple institutions across Australia, working together to conduct research aimed at making urban areas healthier for all living things. Through the tenacious work of incredible individuals in the NESP and CAUL Hub Indigenous Advisories, a position was created within the hub for an Indigenous researcher to undertake work in any aspect of Aboriginal perspectives of biodiversity in urban areas. I jumped at the opportunity to apply.

Thankfully, this role was very open and I was given a huge amount of freedom to chart a course for my work. I chose not to study specific plants in detail, but instead to explore stories about plant knowledge, especially those about plants that are part of the Country on which I was working – Wurundjeri Woiwurrung

Country in Melbourne, or Narrm. I undertook several projects as part of this work, but one of the most successful was a free plant booklet I made to help community groups, individuals and schools connect with Indigenous plant use. It was created to assist them in making indigenous gardens that not only worked to support biodiversity but to educate people about the plant knowledge of diverse Indigenous peoples across Australia.[2]

Through this work, I saw that plants provide a powerful opportunity for learning and connection, opening pathways to reinforce our knowledge, custodianship and continued presence, and to illuminate the all-important specificity of place foundational to our diverse Indigenous cultures. Often it is presumed that Indigenous people, culture and belonging are erased within the urban context. I wanted to counter this assumption by telling stories that highlight that all places in Australia, whether urban or remote, are Country. Through helping people to understand the plants that have grown here, and the many ways Aboriginal people have used them to innovate and continue to hold and to pass on this knowledge today, I assert our place in the present. I try to show people our way of seeing; that no matter how many times you go over her, no matter how violent the alterations, Country still holds her stories and we each, no matter what our culture or race, have a role to play in keeping her healthy and strong. We can nurture Country through what we choose to plant and the respect we show in recognising Aboriginal and Torres Strait Islander knowledge and ways of seeing Country. I have also come to understand the healing potential for Aboriginal community members and for Country enacted through our cultural connection with plants, through the ancestral connection and power of knowing some of the knowledge and stories embedded within our plants, and the need to find ways to reinscribe this connection into our everyday life.

One of my projects was to curate an exhibition, *Emu Sky*, at the University of Melbourne, centring on Indigenous knowledge. Through research, artworks and storytelling, over thirty Aboriginal community members came together as part of this show to tell their stories of Country. One of the artworks, *marrum (overflowing)*, made with a big mob of Aboriginal artists led by Wiradjuri/Kamilaroi artist Dr Jonathan Jones, provided a wonderful learning opportunity for audience members to understand the complexity of our land management practices and plant knowledge, and some of the many ways we have used plants for nutrition, medicine and technologies such as nets and traps. Throughout the exhibition, we asked the audience to consider many of the hidden stories of our people, our culture and our innovation, stories that I will discuss in more depth in Chapter 4.

As an Aboriginal woman I don't like to work on my own. One of my greatest joys is to work with many community members, as I was able to in researching and curating *Emu Sky* and as part of the Indigenous co-author group on the State of the Environment Report 2021, an independent comprehensive national assessment of the state of the environment in Australia released every five years.[3] If things dry up, I will know it is time to look elsewhere. For now, there is a beautiful abundance and I honour those who have placed me within this rich circumstance by working as hard as I can, most especially for my community – bringing people along with me, listening deeply, journeying together.

The work I get to do in telling the stories of our interactions with plants over time is both joyous and unsettling. There are still so many problematic ideas about the efficacy and relevance of Indigenous knowledge systems. Much of the recent heightened interest in our knowledge is highly extractive and damaging for Aboriginal and Torres Strait Islander peoples and our communities, as powerfully evidenced in the alarming lack of benefits we have enjoyed from the explosion of the 'bush food' industry. Our interactions with Country, both today and over time, need further exploration, and Indigenous knowledge will undoubtedly prove a key aspect of the breadth of scientific understanding required to meet the many challenges we collectively face with the escalating impacts of climate change.

I am a relative newcomer to the world of plants, but new and exciting forward pathways keep opening up. Each new avenue has facilitated further work with other mob, in reciprocity and collaboration – an even bigger signpost to keep working hard, to keep going. As I have matured and seen my opportunities grow, I have begun to prioritise empowering my people and their aspirations. I have increasingly looked to my community to guide my direction, asking my elders and others what research they think is important in relation to plants and Country. Overwhelmingly my community is interested in reinvigorating knowledge, and I have tried to aid in this through various research and storytelling projects that speak to our past, present and future, that ask us to imagine what greater empowerment of Indigenous knowledge and, in tandem, greater empowerment of Indigenous communities to manage Country, could look like.

I am quite terrified to be writing this book with Lesley and Michael, two people whom I really admire and who are far ahead of me in their academic achievement and standing. You will notice that my approach is quite different to theirs, because I have much less experience and know much less in this realm, and also because I am not a scientist. I came to study quite late in life and have had lots of different careers. I have an arts degree in History and Australian Indigenous Studies, obtained when I was thirty-five. One day I hope to be seen as a specialist in something, but my work and life journeys have been varied and I feel the richer for it. I approach all of my work from a foundation of relationships and build from there. For me it is a process of listening to community members, trying to work out the stories that they wish to tell and to investigate further, and seeing where this path takes us. One thing I know for sure is that too often the knowledge held in our living communities is underestimated and overlooked, especially in the south-east of Australia, where there is a prevailing, highly damaging, sometimes self-prophesying and utterly false belief that our cultures and knowledges have been 'lost' through the ravages of colonisation. Throughout this book I have tried to share with you the words of Indigenous people who are fighting for their people and Country today, all working to assert our voices and aspirations powerfully in the present.

These days, I work a fair bit with younger mob, and I am always telling them to be proud and to be strong in the different approaches they take as young Aboriginal people, especially within institutional contexts like universities, museums and galleries. These places have only recently found ways to make room for us. Many Aboriginal people can be made to feel out of place and lesser in these institutions, because little is designed to privilege our ways of doing, seeing and being. This makes it all too easy to fall victim to degrading 'imposter' syndrome, because, well, how else are you supposed to feel when nothing fits with the way that you believe is the proper way to do something? When everything we know and do has to be held up to Western science as *the* measure, as if it isn't Johnny-come-lately compared to our bounty. I am trying, through all of my work, to bend non-Indigenous science to be more self-reflexive, to pinpoint opportunities to make room for our knowledge and to engage in truth-telling.

So, while I tell you I'm different to Lesley and Michael, and that they know much more than me, for the young ones I will affirm that isn't a bad thing. It is always worthwhile to let people see the world through a different lens. The First Knowledges series makes room for differing viewpoints by combining Indigenous and non-Indigenous authors, and it has been a great experience to be empowered to tell my stories in my own voice and style. As Aboriginal people, we carry in our essence the impetus and the know-how to take people with us on our journeys. I might not usually be writing books, but I have always been telling stories.

MICHAEL-SHAWN FLETCHER

There are so many reasons I am fascinated with plants. They are the foundation of everything. The architects of life on Earth via their unique ability to convert the Sun's rays into food. It could be the immense feeling I get when in the presence of an organism so large and so ancient that it predates modernity and dwarfs a blue whale three times over. It could also be that despite being largely unable to move, plants have managed to exploit animals (including humans) in so many ways that we – the apparent masters of Earth – might merely be vessels for the world domination of plants. Like wheat, for example, which now gets us to do its evolving for it. Whatever the reason, I love plants so much that my path in life has led me to think and write about them for a living.

I am an Aboriginal man of Wiradjuri descent. I don't know if this is also an inspiration behind my passion for plants. My people, like all people, are completely reliant on plants in almost every facet of life. Aboriginal people all over this vast landmass have used plants for many tens of millennia to both create and read this unique place we now call Australia. So it may be that some part of that deep connection to place, and the plants that make it what it is, has permeated my psyche and led me toward my passion. I guess I'll never know.

I grew up a suburban kid in suburban Melbourne – I called it Melbourne then; I call it Narrm now, its Woiwurrung name. My mother, one of seven Aboriginal children living in commission housing in Sydney's west, decided to make a break for it and move to Narrm after meeting my father. They met at the Wayside Chapel in Kings Cross at an event for underprivileged youth. Mum was fleeing the cycle that was typical of 'half-caste Aborigines' who were the product of the Stolen Generation. A life that drove her to try and scrub her skin white each night after school. This was not a time or a place when you were proud to be Black. Dad, on the other hand, was delinquent. A whitefella squatting in Sydney who was escaping what he felt was a close-minded country community in Bendigo, Victoria, for another existence.

I still live in Narrm today. I should say, I live back in Narrm today. As a kid, my passion for living things saw me outside more than in. I recall loving the time of year when the cicadas would sing. It was still cool and a little damp off the back of the Narrm wet season – what is known in the Kulin Nations' seasonal calendar as Guling, or orchid season, and what most people know as late winter and early spring. The shift into the warmer, drier months is always preceded by that chorus. The fun I had jumping from noise to noise on the nature strips, making them stop, is immeasurable. When the cicadas ventured from their subterranean dens, I'd be waiting. It was like marbles for me. There were plenty of green ones to grab and put on my arm for safe keeping, but I really loved it when I found a golden one or, if I was lucky enough, a black one. Oh, the joy. They were perfect, from their translucent wings to the three weird little dots on their foreheads and their proboscis that allowed them to tap into the energy from the sun via the sweet sap of a tree. And what other tree could it be but the quintessential Australian variety: a eucalypt!

But that was just one event that would mark my year. Before the cicadas, the gorgeous colours and gentle aroma of the silver wattle would tell me it was near the end of winter. This was shortly before I'd have to start ducking for cover from magpies, keen to let me know this was their patch. Then I'd hear the faint squawking of lorikeet babies in the trees I'd pass on my way to school. I'd sit and watch the parents forage on the eucalypt flowers nearby and return to the eucalypt hollows to feed their young ones, before scampering off to school – inevitably late.

What I didn't realise until quite recently was that I was reading Country. I was engaging in science. Building an understanding of the world around me. I started to see young lorikeets emerging before having to fend off magpies while I walked my dog around the local park. I knew that was because different types of eucalypts were being planted that flowered all year round. The same reason we have flying foxes living year-round in Melbourne now: we planted their food.

These connections were floating around in my mind as I progressed from secondary into tertiary education. I was the first in my entire family to do so, having staved off the desire to become a diesel mechanic. I am pretty sure, like most of my brothers and sisters in the academy, I am the only member of my extended family to have completed a degree, let alone go on to do a PhD. Having elected to do science, I aimlessly meandered through subjects in genetics, zoology, human evolution, botany and geology until, on the verge of disenchantment, I was lucky enough to stumble upon geography. Not just any geography, but an attempt to actually engage with and incorporate Aboriginal knowledge and practice into science through plants. This was an awakening for me. All the pieces locked into place. This was my path.

This inspired more radical thoughts. If I had managed to piece together a little of the rhythms of where I live with a few years of observation, what would be the sum result of many tens of thousands of years? How much experimentation would my people have conducted, how much must they have studied? If a keen eye, a keen mind and a five-year journey around Earth on the *Beagle* had inspired one of the most, if not *the* most, important (Western) scientific advances – Charles Darwin's theory of evolution – how much science must reside in the knowledge bank of Aboriginal Australia? What revolutions lie within our knowledge systems?

Since the awakening I experienced, I have never looked back. Sideways perhaps, spending four years living in London, a year or two living and surfing in remote places, but never back. In the process and across multiple continents, I have developed and honed the scientific skills required to establish myself in my chosen field – palaeoecology.

Palaeoecology is the study of how landscapes change through time. It's like time travel. By drilling down deep into a bog, swamp or lake to extract a core sample and analysing what has been trapped in the layers of sediment across time, I get a glimpse of what the place was like in the past. Particles of things like charcoal and pollen hold a wealth of information and are constantly circulating in the air and being deposited on various surfaces. Places like wetlands can store that stuff for tens to millions of years under the right circumstances. Analysing the pollen and charcoal trapped in sediments day after day and year after year over long periods of time, it is possible to reconstruct how landscapes have changed and how factors such as people using fire have influenced those changes. Unsurprisingly, I focus on plants. The architects. My PhD clearly revealed that places we think of as 'wilderness' have actually been constructed by Aboriginal people using fire to manipulate plants.

My PhD research was focused on the south-west Tasmanian 'wilderness' World Heritage area. In it, I demonstrated unequivocally that, rather than being the 'wilderness' that it is renowned as, this landscape was constructed by the palawa people of lutruwita (what we now call Tasmania). So profound was the work that the palawa put into their Country, they were single-handedly responsible for the creation of a completely new ecosystem across the landscape of south-west Tasmania – they expertly applied fire to Country to promote an open grassland-like community where once rainforest had reigned supreme. The scale and nature of this definitively cultural landscape was a firm and empirical rejection of the racist and dehumanising notion of 'terra nullius'. More than this, I was able to use the Western scientific method to prove our agency in creating and maintaining our Country, a system of knowledge production that has historically been used to cast us as sub- or even non-human. A checkmate, of sorts.

My work since has focused on building the expertise and data to form an evidence base that helps substantiate the knowledge of our people on this continent. Data speaks to power in Western systems and I aim to bring irrefutable data to stonewall objections to our knowledge. My lens is plants and they have power. Power to rectify the wrongs committed to my people. Power to stand up to scientific critique. Power to transform how we live on this continent.

This is my contribution to this book.

LESLEY HEAD

My conscious interest in human relationships with plants began when I was studying in the late 1970s and early 1980s. At university, we learnt of various cultural geographers and others who were arguing that the human impact on Earth, particularly through vegetation change, had shaped present-day landscapes and ecosystems much more than had previously been acknowledged. Scholars like archaeologists Rhys Jones and Jack Golson were bringing ethnographic evidence together with the then-limited scientific evidence, to consider how these arguments applied in Australia. Geographers researching long-term vegetation change, like my PhD supervisor Peter Kershaw, were discovering patterns of forest and fire change that forced them to consider possible Aboriginal impacts. Commonplace today, these were controversial ideas at the time, the default explanation being that climate was the key driver of vegetation patterns.

It was an exciting time. The nerdy student could still read everything ever written about the archaeology of Aboriginal Australia and the way the landscape had changed since people arrived here. Before long, each month brought new research papers, making it progressively more difficult to read them all (and it would be impossible now with the flourishing of research in the decades since).[4] Nineteenth-century Northern Hemisphere–centric concepts such as the Stone Age, the ladder of civilisation and hunter-gatherers as primitive crumbled intellectually in the face of new evidence.[5] New concepts – fire-stick farming, domiculture, cultural landscapes,

Country – were put forward and debated to try to understand Indigenous engagement with the environment.

These were challenging and invigorating ideas for the Australian nation. The longevity and achievements of Indigenous peoples were being revealed at sites across the continent, most emblematically at Lake Mungo. These revelations collided with a public context where it was still possible for leading politicians and mining company CEOs to vociferously oppose land rights.[6] Colonisation was seen by the mainstream as something that had happened in the past, and that we should all move on from.

They were also challenging ideas for those of us fighting for the preservation of 'wilderness' areas in Tasmania, northern New South Wales and North Queensland. When I and others blockaded the Franklin River against hydro-electric development in south-west Tasmania, we did so under two contradictory banners: 'Southwest Tasmania World Heritage' and 'You are entering Aboriginal land'.[7] The paradox was that archaeologists had discovered a time before the rainforest wilderness, with a long history of Aboriginal occupation at Kutikina Cave and other sites dating back to the more open vegetation conditions of the last ice age. Fraught relations between archaeologists and Indigenous people in those years have led over time to much greater recognition of Indigenous rights to control their heritage. This is an ongoing process; these relationships have to be continually negotiated in different places and projects.

Another source of challenge was second-wave feminism. Feminist writers and thinkers provided strong critiques of the

'Man the Hunter' trope, stimulating researchers to consider how both women and plants ('Woman the Gatherer') had been erased from the long arc of human history. I can illustrate my plant journey using examples from different projects I have been involved with over that period. We never stop learning and trying to do things better!

Pollen grains

I first came to know the bulrush *Typha*, also known as cumbungi, nearly forty years ago, through its pollen. For the Māori of Aotearoa, the nutritious pollen was a favoured food, mixed with water and steamed to make bread. My approach was via a microscope, piecing together the 8000-year wetland records of south-western Victoria. I was doing this project for my PhD. The cores, containing layers of sediment that accumulated the record of landscape and vegetation change over time, showed that the pollen was often associated with charcoal – direct evidence of wetland burning and land management. This was a time when the first debates about Indigenous impact on vegetation through cultural burning (although the term was not used then) were emerging, but most of the focus was on terrestrial vegetation – rainforests, sclerophyll forests, grasslands.

Burning on wetlands was less discussed but was consistent with ethnographic descriptions, such as those of settler-explorer George Grey: 'the natives bestow a sort of cultivation upon this root, as they frequently burn the leaves of the plant in the dry seasons, in order to improve it'.[8] Burning helped remove dead matter, facilitate harvesting, and stimulate new growth. It also checked the encroachment of *Typha* into open water. I tell a fuller story of cumbungi in Chapter 5, but in brief, when European occupation usurped the Indigenous firing and management regime, *Typha* encroached into a number of wetlands.[9] Europeans had begun turning *Typha* into weeds.

Burning plants

My fascination with precolonial fire regimes and their impacts on plants and the wider landscape led me to seek out situations where Indigenous people were still burning vegetation, and to study the context of their land and resource management. I found the opportunity in the East Kimberley in the late 1980s, where anthropologist Nancy Williams introduced me to the community at Marralam Outstation, a small living area excised from the sprawling Legune Station. There I met Murrinh-patha woman Biddy Simon and Jaminjung woman Polly Wandanga, who became long-term mentors and collaborators.[10] The bigger project that we built up included an archaeological component, led by my husband, Richard Fullagar, a specialist in microscopic analysis of stone artefacts. Over a number of years, we tried to build an understanding of cultural landscapes that had both deep history and continued into the present.[11]

Because the whole area had been heavily colonised by pastoralism, I was told that I wouldn't find people burning, or that if I did it wouldn't be 'real' Aboriginal burning (which in the whitefellas' mind was located in the past, with 'tradition'). But I think no one had really looked, or they had looked and not been interested. Indigenous people were burning the landscape all over the place; indeed, it was sometimes a source of friction with pastoral leaseholders. Those Aboriginal men who worked as stockmen burnt sometimes during their pastoral work, and both men and women burnt often as part of hunting, fishing and gathering trips around Country.[12]

The context at the time was that the colonial frontier was imagined (at least by the colonisers) as a line in time and space that erased or rendered inauthentic all the pre-existing relations that local people had with the land. Colonialism was seen as a single event rather than an ongoing process. The academic context in which I was working, which drew on elements of geography and anthropology, emphasised the recording and preservation of knowledge before the Old People passed away. Although they could not themselves read and write, Biddy and Polly and their elderly relatives were keen for their own knowledge to be recorded, often telling us to 'write that down in the book'. Things have changed in a number of ways, with a much more nuanced understanding of processes of change and continuity in colonial contexts. New generations of Indigenous scholars and writers are telling their own stories for a wider audience, and providing an important reanalysis of colonialism. In northern Australia, colonialism is something of an unfinished project; Euro-Australia still struggles for possession of the land with a frontier mentality, whether in pastoralism, mining or tourism.

Yams

Across this same landscape, Aboriginal women and children regularly look for edible yams in the small rainforest patches dotted across the savannah. With student and later colleague Jenny

Atchison, I accompanied senior women, younger women and children on dozens of regular trips to collect yams. We realised that yam collection involved significant landscape transformations at quite small scales; we came to think of these as yam landscapes.[13] This story is discussed in more detail in Chapter 6. Another smaller-scale transformation was analysed by Jenny in her PhD research. She showed that patterns of fruit tree distribution were closely associated with Aboriginal burning and management; these patterns changed after pastoral colonisation.[14] The terms 'gardens' and 'orchards' never quite fitted, and we were aware of the constraints of imposing English-language categories on complex relationships. The intellectual context in which we were writing was one in which the default assumption was still that landscapes were devoid of human influence until proven otherwise, rather than the assumption that human activities were intimately entwined in the creation of everything in front of us.

Backyards

My interest in Indigenous plant use from the East Kimberley work was confronting, pushing me to consider things I had taken for granted and left unexamined in my own culture. What would it mean to do an ethnobotany of my own society?[15] Why was I going to 'remote' areas to study other people's relationships with plants, when many of the cultural assumptions about settler-colonial relations to plants and the broader environment were completely taken for granted? This reorientation, and the family requirement for fieldwork closer to home at a time when my children were young, led me to the backyard garden as a study site, working with colleague Pat Muir. Would we confirm the notion of settler Australians as destructive and alienated from their environment, or would we find evidence of different kinds of connection and engagement? Our interest in their gardens bemused many of our research participants. It also surprised my family, who knew me as a very half-hearted gardener.

The findings were numerous, showing backyards in Sydney, Wollongong and Alice Springs to be a rich research window onto human–environment relations in urban areas.[16] Contrary to the dominant wisdom that settler Australians were alienated from nature through domesticating and taming it, we found passion and engagement at this micro scale. Households were adapting to sustainability challenges in largely unheralded ways, for example in their conservation of water and their love of birds. We argued that these everyday practices provide lots of cultural resources to address sustainability challenges; many people are ready and willing to do more on this front. Our findings have strong parallels with contemporary climate change debates – the majority of the Australian population is way ahead of the government in their willingness to act. As well as shared attitudes and practices, there were love–hate relationships, and considerable conflict, around trees, lawns, pets, and the desirability of native plants.

There were other project collaborations, on other plants. One team followed wheat into all sorts of nooks and crannies, and in its various forms. Another looked at invasive plants. The East Kimberley story is ongoing across following generations. Jenny Atchison is continuing to work with Biddy Simon, her family and all the Gajerrong-Djarradjarrany Native Title Holders on their Healthy Country plan. Their goals and aspirations include establishing education opportunities, bush food–culture camps, and a ranger group working to identify, map and manage new invasive plants that have recently arrived.

Plants are fundamental to human life – they provide food, clothing, shelter, tools and medicine. Understanding how human life has changed over a variety of timescales requires a range of different methods for analysing interactions between people and plants. But it is also a matter of taking the trouble to look, and to look in ways that assume plants must be part of the picture. The examples in this book draw on a variety of methods: talking to people and recording how they use plants, documenting vegetation patterns, digging into archival sources, analysing plant remains under microscopes. We hope that the book provides insights into the rich variety of Indigenous relations with and knowledge about plants, and ideas for how we can all live with plants in our increasingly volatile future.

2

LOOKING BACK, MOVING FORWARD

LEARNING ON AND THROUGH COUNTRY

Our Country is our university – that's our knowledge centres, our institutes, early childhood centres ... everything.[1]

In this short, powerful assertion, Tati Tati elder Uncle Brendan Kennedy succinctly expresses how we, as the diverse Indigenous peoples of Australia, build knowledge. Although each of the Aboriginal and/or Torres Strait Islander peoples of Australia are distinctly different, there are many aspects of our cultures that can be seen stretching right across our lands. As Uncle Brendan tells us, Country is our foundation. Country holds all knowledge and without Country it is very difficult to pass knowledge on.

Our way of passing on knowledge is rooted in understanding Country as a cultural landscape, comprised of interconnecting relationships. To learn about our world and to continue our culture we must have access to Country, to walk and to sit with our elders, to learn through carefully observing, listening, seeing, talking, smelling and tasting. It is only through being on Country that the complexity of the holistic and integrated nature of our knowledge can be properly realised. This is experiential learning and it puts what you are learning into your body and mind in a way that writing things down, removed from the source or in a classroom, cannot. While books and research are useful, our elders are themselves libraries, and Country too is so, so clever: directing us, demonstrating for us, documenting, providing all that we need to understand and to know. When elders and Country come together our knowledge is activated, flowing from the past, welling up within the present and embedding itself in our futures through continuous application within our communities (Figure 2.1). Today, many communities are building their own projects and resources to document their knowledge, in addition to the continuation of oral traditions.

SPECIFICITY OF PLACE

Our knowledge of Country cannot be applied outside of its context; that is, all knowledge is specific to the Country and community from which it emanates. This can especially be seen in the application of

FIGURE 2.1: Uncle Badger Bates on Country, sharing knowledge about the selection and production of digging sticks.

fire knowledge, where this expertise is highly specific to the plants and other aspects of a particular Country and cannot be reproduced in other contexts without losing its effectiveness, and perhaps causing harm. Often, non-Indigenous science fails to understand the vital importance of the specificity of place and the fact that our communities hold our knowledge, not individuals. It can't be 'taken' or co-opted without losing its essence, its meaning, its veracity, its ultimate value. Often the power of our knowledge resides in the relationships between people that tie it together. When it is moved outside of those relationships, it loses its value. Too often our knowledge is seen as something that can be used or simply added on to existing Western systems to serve the imperatives of others. Once again, Uncle Brendan Kennedy has wise words on the matter: 'We are not a garnish for someone else's meal. We are the main meal, we are the three courses that have always been here, so don't treat us as a garnish.'[2]

It is critical to make clear here that urban areas are Country. In my work looking at Indigenous plant use in the urban context, it has become obvious that many non-Indigenous people perceive the cultural context of Country as 'lost' in urban areas. This is never the case. Country is always alive and always remembers. From the knowing and perspective of Aboriginal and/or Torres Strait Islander peoples, there is no place in Australia that is not spoken for today, does not have belonging to one or more traditional owner groups. It is important to recognise that belonging is multi-faceted. A group may not have formal rights to Country but still has belonging to it through aspects such as ancestral pathways, rights of traversal and kinship relationships. These complex interconnections are often overlooked, but the absence of evidence is never evidence of absence – just because non-Indigenous ways of viewing do not allow something to be seen, it does not mean it is not there. There is a great need to empower and better understand Indigenous communities and perspectives in urban environments. Darug archaeologist and research fellow Maddison Miller explains:

> Cities can give back to Indigenous peoples in a number of different ways. The way in which we plan our cities and the way

> in which we consider our cities can reflect Aboriginal thought and Aboriginal knowledge and Aboriginal principles for caring for Country. If we consider all of the parts of Country, and all of the parts that are important to Aboriginal peoples, we can create better communities. Ones that consider waterways and animal pathways, ones that consider our sacred sites, ones that consider the way in which resources are used and protected and nourish back to the Earth.[3]

Aboriginal people know and understand that Country holds her stories, no matter how violently she is altered, reconfigured or reinscribed. Michael's chapter on his work with the Wurundjeri Woiwurrung community at Bolin Bolin beautifully illustrates this point. Just as our culture is not stagnant, rigid or relegated to the past, so too has Country survived and adapted to accommodate new circumstances and challenges while holding and telegraphing continued cultural belonging.

INDIGENOUS PEOPLE ARE THE FIRST SCIENTISTS

Indigenous science is often perceived as lesser. It is continuously burdened with the baggage of having to be 'proven' against the assumed benchmark of non-Indigenous science. Many scientists today, from archaeologists, geographers and climatologists to biologists, ecologists and ethnobotanists, realise the great value of recognising Indigenous knowledge in their practice. However, this is most common only where they find aspects to support their own scholarship – if Indigenous knowledge is contrary to their study or narrative, it is questioned, derided or erased as myth or not 'real' science.[4]

The core aspects of scientific practice (i.e. 'the scientific method') are observation, experimentation and analysis, as well as the ability to reliably replicate both the tests and their results. When we look at the many examples of plant food preparation across the continent, do we not see these elements in practice? Does it not reflect science to know how to convert an otherwise poisonous substance, often through several distinct processes, into something that has high nutritional value? An excellent example can be seen in the use of nardoo (*Marsilea drummondii*). In order for it to be of use and not detrimental to one's health, it must be roasted, separated carefully from its dark spore cases, then ground and mixed with water to be cooked into a cake.[5] These processes, and the benefits they brought in allowing people to utilise this plant, were discovered through observation and experimentation over time, and the results have been replicated for thousands of years.

The invaders' historical observations tell us that, across this vast continent, Aboriginal and Torres Strait Islander peoples were extremely fit and healthy. Should we not recognise this as hard evidence that the Indigenous peoples of these lands had developed systems of management and subsistence that were and are highly effective? Would their obvious health and vitality not attest to a deep knowledge of nutrition, and of medicinal and technological advancements, especially considering the harsh impediments of many of the landscapes within this continent?

As we move towards a future that promises great environmental challenges, it is timely to enact a process of collective truth-telling in this realm. It will only be when Indigenous and non-Indigenous science find ways of coming together in balance, reciprocity and respect that we will truly harness the many tools at our disposal.

SILENT PARTNERS

The diverse Indigenous peoples of Australia have long been 'silent partners' in the collection and documentation of scientific knowledge. This is highlighted in historian Professor Lynette Russell and biologist Dr Penny Olsen's 2019 book *Australia's First Naturalists: Indigenous Peoples' Contribution to Early Zoology*:

> early European observers left accounts framed by visions of noble savagery and their own assumed superiority, describing Australia's Indigenous people as aimless wanderers who merely lived off the land and made little use of it ... Whether these earliest observers were seeing the Indigenous people as happy savages or struggling brutes, none of them observed the complexity of their social systems or the details of their ways of life.[6]

Botanists and naturalists were essential in the colonising project, and as such their narratives have shaped the dominant perception of this continent's pre- and post-colonial history. They were the 'men of science' charged with 'discovering' the plants that would give the overtaken lands their all-important (Western) economic foundation, through the development of agriculture and various trade industries. But science is not an impartial, impervious collection of 'hard facts'. It deals with measurable quantities, sure, but it is, like all knowledge production, deeply influenced by the societal constructs from within which it is seeded and grows. In this case, the influence of stories of European (namely *white*) superiority – so essential to the colonial project – led to Indigenous knowledge and Indigenous voices being undervalued and afforded little respect, both then and now.

While many early botanists interacted with Indigenous peoples across Australia, these connections were so rarely conducive to respectful engagement. The Indigenous people who accompanied botanists provided vital knowledge of Country needed to find and identify specimens and explain their many uses, as well as word lists and the context of their importance, not only to people but within the larger ecology: for example, their symbiotic relationships with other plants and animals.[7] Despite their centrality to the collection and documentation processes, Aboriginal guides were seen merely as informants, not skilled knowledge holders; resources to be mined and discarded when they no longer served a purpose. Curiously, these early botanists and naturalists are still credited with the discovery of species, completely erasing many millennia of Indigenous peoples' expert knowledge, custodianship, stewardship and innovation related to plants.[8]

TELLING OUR STORIES

The power relationships between Indigenous peoples and collectors were grossly inequitable and there were many compounding problems with the way these 'men of science' both interpreted and appropriated Indigenous knowledge. But these records are still very valuable for Indigenous and non-Indigenous peoples. The reinvigoration of knowledge and practices is highly effective when we draw from a number of sources, including living knowledge still present in communities and the observations, written accounts and illustrations of the invaders. This method of piecing together knowledge from varied sources has been particularly effective in language revival, for example. Utilising multiple voices, we bring together the many pieces of the puzzle in order to make a stronger whole.

Further, not all of these collectors failed to acknowledge the Indigenous peoples whom they encountered, and the knowledges they shared, as explained by Njeri Njeri archaeologist Mark J Grist, reflecting on the work of German zoologist William Blandowski:

> It is important to understand that while we do not know the specific names of the Aboriginal people pictured in the many drawings and written accounts, we do know that they are our people, our ancestors, and that people such as me are their descendants. We do know that the stories we retain came from them. These stories have been handed down to us from one generation to the next. Our oral history connects us to this land and to other lands and their people. As a people, this oral history

> enables us to understand who we are today and enhances our ethos. Blandowski was remarkable because he recorded the Njeri Njeri names for the animals, fish and insects he was collecting. He also published his thanks to my ancestors for all the information and discoveries he was able to make. It was very rare for Europeans to acknowledge Aboriginal people's contributions to gathering knowledge about the 'new land' so this set Blandowski apart from so many other explorers of the period.[9]

Here, Grist also highlights an important aspect to consider – diverse Indigenous peoples have used the same oral and cultural methods to record and to pass information over countless human lifetimes. This consistency must be acknowledged, as it is this uniformity over time, especially in the use of oral traditions, that has allowed information to survive for the longest time imaginable. The oral traditions of the Gunditjmara peoples attesting to the creation of their cultural landscape at Budj Bim are an excellent example of the effectiveness of Indigenous oral traditions. Today, the Gunditjmara people hold oral histories that have been passed down that align with the work of non-Indigenous scientists in understanding the volcanic activity on their Country more than 30,000 years ago, potentially making them the oldest oral traditions in the world.[10] While increasingly celebrated in the arts, oral histories still struggle to be recognised in non-Indigenous science. This is perhaps odd when we consider that there are no written European accounts that have survived tens of thousands of years, yet there are many oral traditions across the Australian continent that span these vast timeframes.

WHO WRITES HISTORY?

While researching for a recent exhibition I curated, centred around an exploration of Indigenous knowledge (*Emu Sky*, University of Melbourne), I did some reading about early European botanists and their scientific study of plants. I came across a study on eucalyptus oil printed in 1883, premised on the assertion that 'the whole medical history of the Eucalyptus oil extends over less than twenty years',[11] a distorted timeframe that completely erases Indigenous scientific experimentation, use and application. While the historical modes of scientific study were somewhat amusing, it was difficult not to feel disillusioned that such a whacky cavalcade was considered scientific practice when taking instruction from the Indigenous peoples who have utilised eucalyptus for many millennia was somehow considered lesser:

> Here it is remarkable to note that Professor Schulz himself took the oil, purified and oxygenized by his method, as much as two scruples in a single dose without inconvenience; and even when increasing the dose to three scruples (of course diluted) he merely felt depression, but nothing abnormal. Moreover, he observed, after taking the refined and oxygenated oil, that the urine, particularly after warming, assumed the odour of violets.[12]

Well, I'm unsurprised that imbibing eucalyptus oil brought Professor Shultz down. Throughout the course of this research, Dr Shultz documented some lethal experiments on an assortment of animals variously injected with or otherwise exposed to eucalyptus oil, as well as recording some particularly nasty after-effects from weeks of experimenting using his own body. Of course, self-experimentation was more common in the past, but these grim tales make a reader wish (even more) that he had just engaged with any Indigenous person, who would have saved a lot of discomfort and a lot of innocent animals by stating, in one way or another, 'Mate, external use *only*!'

This example perhaps gives pause to think about what does and does not constitute accepted scientific practice, and how some of the exclusionary barriers of this realm have not adequately changed over time despite accepted practices significantly evolving. The integrity of our knowledges continues to be scrutinised to this day, while the historical foundations of modern-day scientific practice are explained as simply 'steps in the right direction'. The transmission of knowledge through our living communities to create highly effective economies, cultural practices, systems of management, technologies and nutritional and medicinal discoveries is not, it seems, evidence enough. Ultimately, the lack of care, respect, attention and value given to Indigenous knowledge and the failure to empower it since the beginning of colonisation shows a deep-seated bias that is, given the available evidence, entirely *unscientific*.

ERASURE

In order to understand the historical and contemporary relationship between Indigenous and non-Indigenous science, we need to consider their earliest engagements, particularly the ways in which Western science has actively participated in the oppression of Australia's First Peoples and our knowledges. This unpacking also allows us to understand some of the thinking behind this exclusion and erasure, especially as we work to identify the narratives that continue to negate Indigenous knowledge and cause silences and harm today.

Having (still quite recently) moved away from the myth of Aboriginal people as 'aimless wanderers', we are beginning to come to a more complex and layered understanding of the interactions with Country, and the deep knowledge underpinning it, practised by diverse Indigenous peoples over time. While some may argue over the words used – such as agriculture vs hunting and gathering[13] – what matters more is that heightened regard is being shown, and we are starting to see much more nuanced research and focus on the land management practices and unique understanding of Country held by Australia's Indigenous peoples.[14] Further, these efforts are helping to dispatch untruths in understandings of our economies and management practices.[15] For example, the central role of women in food procurement – the majority of the daily meals of most Indigenous communities traditionally come from foods, particularly plant foods, gathered by women – has been distorted by prevalent assertions that revolve around men hunting game.[16]

These narratives have been aided by the imperative to show us as savage. This can be seen in the way our cultures have been collected and displayed over time in museums, where our agricultural tools and other items that show complex innovation, knowledge and management have been barely recognised, while our weapons often take centre stage.[17] These same patterns of reductionism are clear throughout the available historical literature. The vast majority of accounts tell us that diverse Indigenous communities were incorrectly understood as homogenous, as one large indistinct group. There are many accounts that cruelly assert that Australia's First Peoples were seen as without humanity, intelligence, or complex social and ecological systems of management.

Despite an increase in opportunities for engagement with institutions, they remain largely unsafe spaces for Indigenous peoples around the world. Almost always making up a tiny minority of most institutions' staff, we are too often lone boundary riders across sharp and often dangerous borders existing between all-powerful institutional structures. These places have been made to uphold European values and ways of doing, thinking and being, and have been established without us, and specifically to exclude us. The continued exclusion of our peoples, ways of doing (including our pedagogies) and knowledge from the 'mainstream' exemplify the ongoing structures of colonisation. While our knowledges are increasingly recognised as valid and highly valuable, it is still rare that our ways of doing are empowered or enacted. All mainstream systems of knowledge production in Australia continue to privilege non-Indigenous ways of seeing and doing. Nyikina Warrwa academic Dr Anne Poelina works tirelessly to forefront not only Indigenous knowledge but also Indigenous pedagogies, demonstrated in her inclusion of the Mardoowarra/Fitzroy River as co-author in a recent academic publication.[18] But this type of recognition and promotion of our worldview is still too rarely seen. While increasingly visible in public life, our pedagogies remain overwhelmingly unrecognised in Western systems. The ingenuity and usefulness of Indigenous knowledges (and their custodians) are appreciated in an academic sense, yet they remain as 'other', rarely integrated into policy or practice, which continues to honour the lie that our knowledge and ways of doing are lesser.[19]

Compounding these false narratives about our past and present cultures, many institutions provide stark reminders of the dual histories of 'dying race' theory and the theft of ancestral remains. Built upon the false notion that Aboriginal and Torres Strait Islander peoples are inferior and represent an early stage in the evolution of humankind, the pseudoscientific 'dying race' theory heavily influenced public policy in the early 20th century.[20] Dismissing Indigenous peoples as being predisposed to genetic weakness so pronounced as to be incapable of survival as a race,[21] its mark can be seen in the thinking and 'evidence' used to enact the devastating policies of child removal in Australia that created the Stolen Generations.[22] Sadly, the idea that we are racially inferior is not a thing of the past. Perhaps the way it is expressed today is somewhat less overt, but it is no less damaging.

Of course, no one now can credibly argue outright that our vibrant Indigenous cultures are dead. However, if you perpetuate the idea that a culture has been devastated and irretrievably lost, you insist that there is nothing left to look for, no need to shine a light because there is nothing to be revealed, researched, reinvigorated, celebrated or empowered. When narratives of loss and inauthenticity are accepted then our communities and knowledge holders become drastically under-resourced.

I have too often been contacted by young scholars wanting to do plant research who have told me that 'experts' have advised they will have to go up north to 'traditional people' or they will find nothing of worth. This takes keen young researchers away from projects that could bring much value to Aboriginal communities across south-eastern Australia. Perhaps part of the problem is the 'great white explorer' mentality that can sometimes be seen in research, where there is a desire to be the 'discoverer' of 'new' untapped information with the problematic idea that only those groups living what is perceived by outsiders to be a more 'traditional' way of life hold information waiting to be 'discovered'.

Dr Cressida Fforde beautifully explains notions of 'authenticity' in this context (although written some years ago, this remains a problem with perceptions of authenticity today):

> Although Australian legislation no longer employs definitions of Aboriginal people based on genetic inheritance, it has been argued that an assumed association between biology and culture continues to exist. There is a frequent popular academic (and not exclusively non-Aboriginal (see Myers 1994:690)) division made between 'traditional' and 'non traditional' or 'urban' Aborigines, and the perception that only the former are somehow 'real' Aborigines. Such perceptions deny Aboriginality to those who do not exhibit what is perceived to be 'pristine' (i.e. pre-contact) Aboriginal culture.[23]

Historical documentation and these patterns of thought and the erasures they enable are expertly examined in Uncle Bruce Pascoe's *Dark Emu*, but we each can and should interrogate damaging assertions of irretrievable loss, as well as continuing to question erasures in mainstream science and history, their roots and their underlying logics.

EMPOWERING COMMUNITIES TO CHANGE THE STORY

Indigenous communities across the south-east and beyond are doing powerful work to foreground their cultures and knowledge,[24] especially in relation to plants. Many traditional owner groups are undertaking impressive revegetation and traditional food projects and working in collaboration with researchers to strengthen and tell the stories of Country and ensure their longevity. There is growing recognition that the way Country has been managed since the beginning of colonisation is unsustainable and that ignoring Indigenous expertise is to the detriment of all. Many in our society are starting to question the power imbalances that have existed for too long and the damage they have caused and continue to cause. Today, Indigenous peoples are demanding to be heard, to lead projects that educate others about the importance of our knowledge of Country. An excellent example can be seen in the inception of a collaborative biodiversity project led by the Dja Dja Wurrung Clans Aboriginal Corporation (DDWCAC):

In 2019, DDWCAC and Parks Victoria started a pilot project called Walking Together – Balak Kalik Manya (Community: People meeting with many hands). The project is to deliver and implement site specific management plans for two areas near Dja Dja Wurrung townships, with the aim of further connecting community with nature, whilst protecting and improving biodiversity. The two areas are Kalimna Park within Castlemaine Diggings National Heritage Park, and Wildflower Drive within the Greater Bendigo National Park. Both areas have minimal onground management, and there is a desire from Parks Victoria and Dja Dja Wurrung for more intensive management to emphasize culturally appropriate management of land.[25]

Urban areas can be seen as the front line as we consider how we might combat climate change and the work we need to undertake to reconfigure the lack of custodianship and care that has caused such degradation and devastation over a relatively short period of time. Urban areas are intrinsically connected to our environment as a whole and contribute disproportionately to issues in the wider environment. Put simply, this is because of the high population density in these areas, meaning pressures such as energy use, pollution and threats to biodiversity caused by the development of land occur at a higher rate compared to non-urban areas. The heightened circumstances of the environmental challenges we face call for new modes of partnership to be forged, new ways of bringing together knowledge and a reckoning with extractive and insincere partnerships that must be discarded. The lack of recognition of traditional owners and the incorporation of their cultural values in mainstream planning in urban areas over time can be seen in light of the issues of authenticity I have spoken of.

Many south-eastern communities, and particularly urban Indigenous communities, are perceived by outsiders as inauthentic, as having lost their culture, and this has effected a silencing of Indigenous cultural values and knowledge in the planning and management of urban landscapes. While it is true we have suffered greatly under the ongoing harm of colonisation, our urban communities and our cultural belonging remain proud and strong, and no person has any right to deny or erode our powerful place in the present. Traditional owners in urban areas are slowly beginning to find their rightful position as respected knowledge holders, evidenced, for example, in recent cultural burning projects in Melbourne and Adelaide with Wurundjeri and Kaurna traditional owners,[26] as well as in cultural mapping projects in Perth[27] and in the passing of significant legislation in Victoria that foregrounds traditional owner perspectives of the Birrarung/Yarra (*Yarra River Protection (Wilip-gin Birrarung murron) Act 2017*).[28]

Many exciting projects are emerging that provide more opportunities for us to tell our stories our way and, most importantly, to get people back on Country to fulfil custodial obligations for the reciprocal benefit of all. And on a local scale, Indigenous peoples around the country are continuing plant knowledge practices within their communities, working to reinvigorate knowledge that has been impeded through the ongoing pressures of colonisation. The challenge is ensuring that Indigenous and non-Indigenous people find ways of working together that ensure mutual respect and benefit sharing. We are finding ways to tell a collective truth, to dispel many of the damaging myths that continue to cause harm today. Our stories will be richer when we find ways of telling them in true partnership, no matter how messy or complicated. Together we may light a path forward that can lead us to where we need to go – to bring necessary reparations to people and to Country.

3

BOLIN BOLIN

Landscapes are, in large part, defined by their plants. At the global scale, plants define Earth's biomes (environments characterised by particular climates, plants and wildlife, such as grasslands, forests, deserts, etc.). At the continental scale, they define ecoregions and ecosystems. Plants are at the core of life as we know it. They fix the almost inexhaustible energy that thunders down to Earth's surface from the Sun into a form that makes it accessible to all life on Earth. They are truly remarkable. Without plants, we would have no food or oxygen. In many ways, the success of our species depends on a successful relationship with plants. It is through plants that we humans manipulate and shape the world around us. It is through plants that our people care for Country. It is through plants that we made this continent. And it is through plants that we will rescue Country from its mishandling since the British invasion.

THE BOLIN BOLIN STORY

To begin to get a picture of the past, we need look no further than the city where I have lived most of my life: Narrm (now known as Melbourne). The Narrm of today is a radically different landscape from what it was under the custodianship and management of the Wurundjeri and Boon Wurrung/Bunurong. Where today there is a concrete jungle surrounded by what feels like an interminable urban sprawl, there was once a landscape of vast wetlands and grassy plains focused around where Birrarung (now the Yarra River) met the sea. This whole region was a high-production environment where fresh water, salt water and land met. A system in which the Wurundjeri and Boon Wurrung/Bunurong managed more than 150 species of grass, twenty-four species of rush and sedge wetland plants, more than 180 species of birds (including culturally important species such as magpie geese, brolgas, swans, ducks, emus and bustards), more than twenty species of fish and eels, and two species of freshwater crayfish. It is no surprise, then, that within five years of first 'discovering' Narrm, more than 20,000 British and other settlers had flooded to the area, along with more than 700,000 sheep.[1] Like much of this continent, the British capitalised on what was a deliberately curated and managed landscape, while giving absolutely no credit to the people who created it.

Further up Birrarung lies a series of billabongs (also known as oxbow lakes) of significant importance to Wurundjeri people to this very day. Billabongs form when sinuous rivers on broad, flat floodplains suddenly change course and leave behind an often horseshoe-shaped lake that was once part of the river. These wetlands continuously form and slowly fill up, eventually becoming part of the floodplain. One of these sites, Bolin Bolin Billabong (Figure 3.1), was the meeting place for hundreds of Wurundjeri for months at a time, where they would feast on eels, fish and other products of their careful curation of their Country. So important is Bolin Bolin Billabong that it was the site chosen by the Wurundjeri as their preferred place to settle while attempting to negotiate an agreement with the British authorities after invasion. This request was rejected, as the British desired the productive land for themselves.

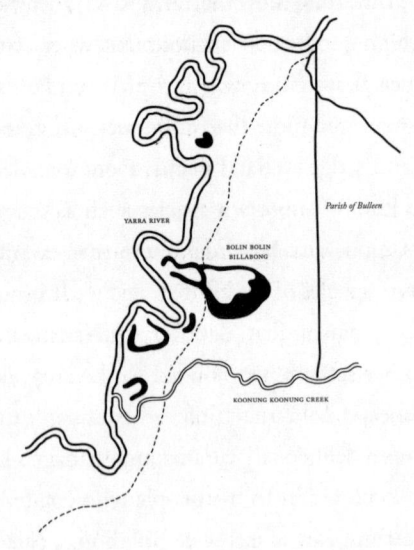

FIGURE 3.1: Map of Bolin Bolin Billabong and surrounds, drawn in 1837 by the surveyor Robert Hoddle.

Today, Bolin Bolin feels like a forgotten ruin in the inner suburb of Bulleen, a suburb which derives its name from the billabong itself. A busy thoroughfare of trucks and traffic bustles past the site continuously. The water has a bright orange tinge and the mud a deep black colour that reflects a wetland heavily polluted and starved of care. Giant branching *beal*, or river red gums, dot the shore, relics of a former time, when Country was more open. Now, these giant bicentenarians are crowded underneath by shrubs and young trees jostling for space and light. This new generation will grow tall and straight, not having the luxury to spread wide in open Country, free from competition. Further towards Birrarung, the sound of the passing trucks fades and kookaburra laughs can be heard, kangaroos glimpsed among the dense weed-infested shrubbery. Happy strollers frequent the paths, walking their dogs in what they no doubt think is the bush, much as it was before the city grew too big.

While Bolin Bolin is forgotten to most, it is not forgotten to the traditional custodians. To them, it is as important as ever. It is through this place that I first collaborated with the Wurundjeri. I had previously learnt how important this and other remnant wetland sites were. I had also heard of how they were working towards rewatering the billabong to improve its health, as it is currently starved of connection to Birrarung by the heavy regulation of water in the Birrarung catchment – 'Starved from its mother', as Wurundjeri elder Uncle Dave Wandin eloquently and accurately puts it.[2] I offered my tools to the Wurundjeri to see what secrets of the past were stored in the sediments of Bolin Bolin. After all, this place was a veritable supermarket. A jewel in the landscape. After talking with Wurundjeri on Country at great length, it struck me how important this place is and how the overgrown and polluted state of the site was like a wound for them.

READING HISTORY

My work lets me step back through time in any given place. Wetlands are a portal to the past. Think about all the stuff you breathe in: smoke, dust, pollen, chemicals, bugs, bacteria, and so on. They are all released into the atmosphere and mixed around until they eventually settle on the ground, on your skin, on top of your fridge, under the couch … you name it, if an object sits around long enough, it will start collecting atmospheric information. Ever been out to the garage, poked around and blown the dust off that old wooden chest, a collection of *National Geographic* magazines or a vinyl record? You, in effect, are removing information that has collected day after day, year after year, waiting to be read. Given the right tools, you can read that book. That natural archive.

Natural archives abound out in the world. They include the sea floor, stalactites and stalagmites, tree rings, bogs and, last but certainly not least, lakes. Lakes are ideal sources of environmental data, gathering information from the atmosphere, depositing it deep underwater and preserving it for thousands or even millions of years. This is a story waiting to be told. A story that gives us a glimpse into the past. A window through time.

So, on a boiling hot day in January 2019, and with a crowd of keenly interested Wurundjeri, I set to work extracting a sediment core from Bolin Bolin Billabong. This involved constructing a pontoon, which was then floated to the centre of the billabong. From this platform, I could carefully drill down into the bed of the billabong, taking a sample of the layers of sediment from beneath the water. I am always somewhat nervous when coring. Not only because everything I do in the laboratory after this point is dependent on getting it right at this stage, but because I know I am unearthing someone else's story. I know, when working on this continent, that Country was cared for and that I will be, in some way, glimpsing into that care and into the lives of the people who cared for that Country. This time, it felt like I had the weight of the past and the present Wurundjeri with me.

Around six hours – and some six meters of sediment and river gravel – later and the job was done. I had in my care the entire history of Bolin Bolin Billabong, from when the river changed course to make the billabong through to that hot January day. After two years of pandemic interruptions and tedious, time-consuming lab work, Bolin Bolin's secrets were revealed. And never could I have imagined the story it would tell. The power it would provide the Wurundjeri and all Aboriginal people. The power to be believed. The data produced by the very Western scientific system that has disempowered our people for centuries. Data to back up the knowledge of our people. That this is, and always will be, Aboriginal land. True to form, Bolin Bolin is indeed a special place.

A YOUNG BILLABONG

Bolin Bolin Billabong is quite young, having formed in the mid to late 1700s. This could come as somewhat of a surprise until you realise that rivers are always evolving. The present-day Bolin Bolin Billabong is one of more than fifty billabong features still visible along this section of Birrarung and lies adjacent to a larger and more ancient billabong under what is now the playing fields of an exclusive private school. Immediately after forming, the vegetation around Bolin Bolin Billabong was a fern-rich rainforest dominated by myrtle beech (*Nothofagus cunninghamii*). That's right. Rainforest. Right within the lands of the Wurundjeri, there was rainforest. The closest rainforest to Bolin Bolin today is 30 kilometres away in the mountains to the east of Narrm, whose surrounding environment is considered unsuitable for this rainforest type today. In the past, there was enough rainforest growing along Birrarung that when this billabong formed, the rainforest was able to move in and capture the site almost immediately. Another interesting feature of this period is the absence of any flooding for more than twenty years while the rainforest surrounded the site.

Rainforest in Victoria is now almost exclusively restricted to the high country, in areas protected from fire. It seems that, unlike the past 180 years under British control, the Wurundjeri were able to manage fire on Country in a way that allowed rainforest to persist near sea level in Narrm. This should not be a surprise, given the ample evidence from across the continent of Aboriginal people actively protecting fire-sensitive plants via our detailed and

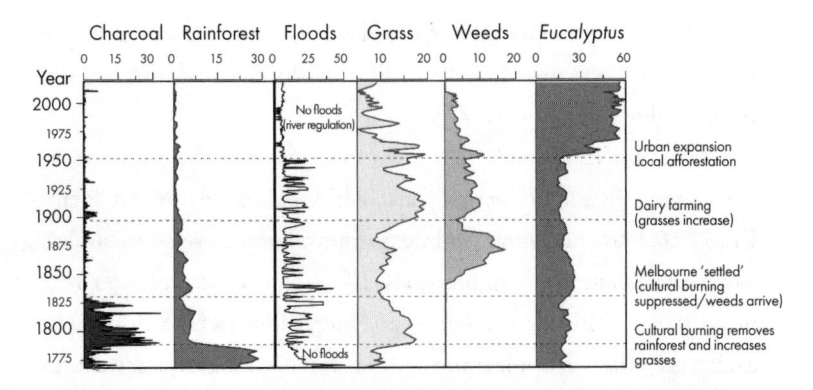

FIGURE 3.2: A summary diagram of the palaeoecological data from Bolin Bolin Billabong. The data include charcoal fragments as an indicator of fire, pollen types as indicators of plant types and material deposited by Birrarung flood events. These data are shown as a percentage of a base sum of terrestrial plant pollen types. Key points through time are shown on the right.

sophisticated manipulation of fire and its influence on the structure and connectivity of fuel. It is this fine-scale management that makes Aboriginal-managed landscapes more diverse than unmanaged landscapes. It is the detailed knowledge of plants, landscapes and fire that this continent was built with.

What happened next in the story of Bolin Bolin Billabong is, in my humble opinion, the most noteworthy and important offering from the sediments I gathered. Within twenty years the Wurundjeri had removed the rainforest surrounding the site with fire. It is clear that there was still rainforest nearby, but it was deliberately and systematically removed from around Bolin Bolin Billabong. Immediately following the removal of the rainforest, Bolin Bolin Billabong experienced regular flooding from Birrarung.

The opening up of the vegetation at the site, the manipulation of what plants were growing there, allowed floodwaters into the site more readily (see Figure 3.2).

This act is a powerful one – and a powerful statement. Aboriginal people deliberately altered the landscape to suit them. Cool temperate rainforests, while containing some useful resources, are generally poor in commodities. Billabongs connected to rivers via regular flooding, however, are incredibly rich in resources. Eels and other fish need connection to rivers and waterways to complete their life cycles, while regular flooding brings nutrients and sediments that are critical for the health of the entire landscape. The speed at which the Wurundjeri acted to convert the new billabong to their preferred state indicates a system of management that was reflexive and reactive to the constant addition and removal of billabongs within the floodplains of Birrarung; a system in which people were continually working to maintain productive and predictable Country.

We have little understanding of how the arrival of our people influenced the landscape of this continent. This is in part because it was so long ago that there is very little oral or physical evidence to draw on. We don't even know when our ancestors arrived here, only that we have been here long enough for more than 3000 generations of people to have lived and worked on Country. Long enough for 'forever'. We do, however, know that we have worked with fire through much of our traceable history, and that humans have used fire for more than 1.5 million years. A large part of this time has been devoted to using fire deliberately to modify landscapes by inserting more grass. Today, many Indigenous and local people continue to use fire for this exact purpose.

GRASSLANDS

Grass is at the heart of the story of the country now called Australia. If one thing can be said to bind all human endeavours prior to the industrial revolution (and likely since), it is that humans depend on grass. Increasing the grassiness of landscapes is at the core of most of our landscape management. Grasses produce the grains on which the world is almost entirely reliant, and they form the food of almost all the animals we consume. And as evidenced at Bolin Bolin Billabong, a shift to a more grassy landscape can also activate and help bring renewed life to local waterways. This deliberate manipulation of Country using grass has occurred throughout human history, and for tens of thousands of years on this continent.

Thus the first impressions white people had of this land were entirely based on plants, and on grass particularly. The south-east parts of our country were so reminiscent of the manicured English countryside that almost every written account was full of superlatives describing the scene as a 'gentleman's park' containing some of the 'finest meadows in the world'.[3] Boon Wurrung/Bunurong Country, surrounding what is now known as Port Phillip Bay, where I have lived most of my life, was described as 'enchantingly beautiful' with 'extensive rich plains ... having the appearance of an immense park'.[4] Yet, unlike the 'gentleman's parks' in England, which were intensely curated estates, our Country was deemed to be 'natural'. In that same passage, the author concludes that Boon Wurrung/Bunurong Country was 'a lovely picture of what is *evidently intended by Nature* to be one of the richest pastoral communities in the world'[5] (emphasis added). Of course, this was no happy accident of nature. The Indigenous peoples of this continent were working the land, but in a very different way to what the British could understand.

In the (apparent) absence of evidence, humans fill in the gaps using their understanding. This understanding is wrought from our ontology or 'worldview'. The British saw our lands, but they did not see us. Having long since lost their connection to fire on the journey towards their idiosyncratic type of agrarian system, developed on incredibly fertile soils in a remarkably stable climate, they did not have the skills required to read our Country managed in this way. They could not see past the missing fences and farmhouses, or the absence of familiar farming equipment. Worse, however, is the fact that these biases were compounded with other even more harmful and pernicious misconceptions: primarily, deep-seated and deeply flawed beliefs in European racial superiority.

TERRA NULLIUS

One of the key prisms through which Aboriginal people have been denied their humanity is through the idea that we did not own or work our land. It is the central tenet behind terra nullius – the concept that ownership by seizure of a thing no one owns is legitimate – a law that paved the way for one of the grandest acts of larceny ever committed: the theft of an entire continent. This myth continues to pervade all levels of society to this day, so let us interrogate this notion a little further.

That Aboriginal people are keen observers of their world is undeniable. The complex interweaving of the living and non-living, human and non-human worlds reflected in the seasonal calendars produced by different communities attests to a deep understanding of Country. These bodies of work reveal an encyclopedic knowledge of plants, animals, climate, landforms, and the position of people within Country. Rather than being fixed in time, 'seasons' are defined by events, such as plants flowering and animal migrations. This flexibility is of purposeful design. A design fit for a variable place in which climate is not regular. It allows for appropriate resource management, with an approach based on listening to Country, not by fixing a date in the calendar. While the incredible detail contained within these calendars is garnered from long observation and deep understanding of Country, calendars are not able to speak to our role designing our Country.

Is there evidence for landscape design? Of course there is. The large-scale stone fish traps found across the continent attest to our ability to design and alter the hydrology of landscapes into aquaculture systems. A process that has been occurring for at least 6000 years and one that is employed across the continent. These aquaculture systems were productive enough to support large permanent and semi-permanent villages. Recognition of these as intentional structures within the mainstream was hard fought. All manner of challenges were presented and overcome to unequivocally prove that the stone fish and eel traps were deliberate features surrounded by village sites.

One of the more ludicrous obstacles was that stone hut depressions were actually fallen tree stumps that had unearthed a rim of stones. Why these unearthed stones should then have archaeological material and fireplaces was beside the point. This narrative did not fit the foundations on which this continent was built. These challenges were overcome by merging Aboriginal knowledge with Western science. By listening to the traditional custodians and being led by their knowledge, scientists were able to understand where to look and what to look for. So successful has this Aboriginal-led initiative been that the Gunditjmara have now had their landscape recognised as a World Heritage area. A key aspect of this successful bid was proving the longevity of their systems of aquaculture through the combination of cultural knowledge, oral traditions and Western science. Ultimately, they were able to demonstrate that they are among the first, if not *the* first, people to practise aquaculture on Earth![6] But still, we aren't allowed to call ourselves 'farmers' – that's a stretch too far.

We were parasites. Noble savages living in balance and harmony. Deploying some modicum of skill to eke out a 'savage' and 'beastly' existence from what nature offered us. We had no agency, were not *capable* of agency. We were, at best, the missing link between apes and Europeans in what Charles Darwin would eventually coin 'evolution by natural selection'. We were thus naturally inferior. Incapable of producing such a bountiful landscape. It was a scientific fact, and nothing is more irrefutable than a scientific fact. This certainty paved the way for the theft of our land, for our graves to be robbed, for our children to be stolen. All crimes of the highest order in British society, excused by the infallibility of science.

While the most overtly heinous of these crimes have abated, we are still dispossessed of our land, our children are still being taken away and our agency as the architects of our Country, of this continent, is still routinely denied. Further, white people still speak on our behalf without appropriate consultation, and debates about what and who we are is the sport of intellectual elites, politicians and shock jocks who toy with our lives with no concern for our voice.

RESTORING BALANCE

How do we reset the balance? Can we use the scientific tools employed to cast us in these fictitious roles to shed light on the truth? Yes, we can, but it requires an acknowledgement of what science is and what science is not. All cultures use what is now called science to sustain themselves and understand the world. It is an innately human pursuit, to observe and experiment in an attempt to make better predictions and achieve desired outcomes. But despite popular misconceptions, science is demonstrably not 'objective'. As Zena has outlined, all observers (i.e. scientists) are biased by their cultural viewpoint. All of us bring our worldview into the pursuit of science.

Having established those two critical points, we are forced to recognise that attempts to investigate cross-cultural topics using science – such as investigating the origin of a foreign landscape – are doomed to fail if the problem is not considered from multiple viewpoints. The most obvious example I can think of is to consider the ship that sailed into Port Phillip Bay in 1836, from which the observations of Boon Wurrung/Bunurong Country that I elaborated on earlier were made. What if the ship was sailing up the River Thames in England or the Seine in France? Would the observer make the prediction that the banks of these rivers were 'as intended by nature'?[7] I would think that same observer would readily recognise the fields and the influence of humans in those landscapes. This is a more-than-obvious caricature of observation bias, but it underscores an important point.

I will extend the example through the lens of my own scientific field, palaeoecology. Framing a landscape as 'natural' places the burden of proof on proving unequivocally that people created the landscape. If I was to frame the same landscape as 'cultural', the burden of proof now lies with proving beyond doubt that humans did not create the landscape. You might say, 'But yes, it is obvious that the banks of the River Thames were a cultural landscape – look at all the fences, farmhouses and livestock – whereas the land around Port Phillip Bay had none of these and looked natural.' Herein lies the bias, experience limiting one's perception. To whom is it obvious? Would a Gunditjmara observer in this Boon Wurrung/Bunurong landscape make the same observations and predictions, or would they recognise the signs that this was managed Country? Would they see the plants as cultivated? The kangaroos as livestock? Observation bias dictates the predictions (or hypotheses) we make and the questions we ask. These, in turn, determine the experiments we undertake and, most critically, where the burden of proof lies when testing predictions.

Merely turning scientific approaches, such as the one I employ in my day job, towards questions on Aboriginal lifeways is not enough to break the cycle of what anthropologist Deborah Bird Rose calls 'a hall of mirrors':[8] the inability of Western scientists to understand that their cultural bias defines the answers they derive from their scientific pursuits, thus reflecting their own biases back at them. Aboriginal people must be allowed to set the paradigm, ask the questions, interpret the data. This is even more pressing in a world in which most decisions are made based on data. Data speaks to power. A power we have always been denied when non-Aboriginal people speak on our behalf or in our place.

A VOICE

Who controls the past controls the future: who controls the present controls the past.[9]

This George Orwell quote has no truer fit than within Australia. Aboriginal people have been and continue to be governed by the first impressions of Country by the British. By the myth of terra nullius. We have been cast in many roles by non-Aboriginal society. Rarely have Aboriginal people been allowed to speak. We are the subject or the object, never the narrator.

Our people have long battled for recognition in this country. This is perhaps best reflected in the fact that terra nullius was still enshrined in Australian law up until 1992. Our way of life and our relationship to Country have been debated by non-Aboriginal academics, intellectuals and laypeople since the British began their invasion of our lands more than 250 years ago. We have been pulled apart, literally and metaphorically, dismissed as 'noble savages', 'intelligent parasites' and the 'missing link' between humans and apes. Indeed, a dominant belief in anthropology more than 150 years after the British invasion was that our people were merely 'subservient to nature' and capable of no more than becoming 'skilled in observing her ways, and in gathering and catching her gifts of vegetable products, marsupials, reptiles, fish, birds and insects'.[10]

These views have never really disappeared. Until recently, Aboriginal voices, if considered at all, were almost always conveyed through non-Aboriginal people. Any Black people who tried to insert their voice and contest the stories being told about us were rapidly cast as unruly antagonists and spurned by intellectual elites for their lack of formal education and training. If these same unruly and uneducated Blacks happened to be 'half-caste Aborigines', then they were dragged through disgusting identity politics and forced into the crevice between being too white to be Black and too Black to be white.[11] If we attempt to get educated so we can make a difference, we are cast as 'colonised' Aborigines who are disconnected from culture. We still see this same tried and true approach today when Aboriginal voices say something that those with power dislike. While there is a change in the wind of sorts, it is hard to take seriously the majority of efforts to 'incorporate Indigenous knowledge' in academic and other pursuits when you understand who fashioned the foundations of our modern institutions and how zealously some of these ideas are still clung to. Nevertheless, we are regaining our voice and we are infiltrating Western institutions. We are speaking your language and playing your game. This book series is a reflection of this.

The volumes of writing on who we are and what we do by non-Aboriginal people continue to accumulate. The same tropes are recycled time and again, and we are forever locked into place and out of the conversation. Examples include *The Future Eaters* by Tim Flannery, where events from other continents are used to cast us as environmental vandals who wiped out all the big animals, which forced us to then burn the resulting excess plant growth, despite insubstantial evidence to support this claim; *The Pure State of Nature* by David Horton, a scientist who spent most of his life working for an Aboriginal institution and whose work casts us as mere parrots of nature who did nothing but mimic what the environment laid out for us; and the recent contribution, *Farmers or Hunter-Gatherers? The Dark Emu Debate*, led by Peter Sutton, who embarked upon a semantic crusade to deny an Aboriginal voice that is essentially based on gatekeeping around the meaning of words like 'farming' and 'agriculture', and the shortcomings of the English language to adequately describe what we are. All intellectual white men who think they know more about us than we do. All recycling the myths and misconceptions that lie at the foundations of modern Australia.

BACK TO BOLIN BOLIN

So, how does all of this relate to Bolin Bolin Billabong? The active management of Bolin Bolin by the Wurundjeri created and cared for a more diverse set of wetland and terrestrial plants than under British occupation. A pattern we see spanning this vast continent: Country is healthier under our management. Not because we mimicked nature, but because we created, shaped and cared for our Country. Our land. This fact is well known by Aboriginal people, and will hopefully be acknowledged by the rest of Australia now that it has been confirmed by the analysis of Bolin Bolin Billabong using the techniques of Western science. This research has taught once again that knowledge of Country resides in the people who have lived on it 'forever'.

This lesson cannot be overstated. Aboriginal people intentionally changed Country to suit their needs. We created our landscapes, landscapes that the British adored so much that they declared them the 'finest in the world'.[12] These are not natural landscapes. They are part of deliberately curated Country. Country that was managed to suit the needs of our people. This simple scientific reality strikes down the notion that Aboriginal people were mere 'parasites' or effected 'little if any change' in the Australian landscape. It strikes at the very core of terra nullius and reveals the grand larceny that modern Australia is founded on. It rewrites the fictitious history upon which this country is built. A story viewed through plants.

4

ABUNDANCE

The truth resides in places that are invisible. Once you are aware that there is a different world out of sight, you will be living in a different way.[1]

The photograph overleaf (Figure 4.1) is one of my favourite images of all time for a multitude of reasons, not just because it shows expert skills in harnessing the many uses of plants. First and foremost this photo is so loved because it depicts my people, the Barkandji.[2] Barkandji Country is situated in western New South Wales and stretches from Bourke down the Barka/Darling River to where it joins the Murray River and also includes large sections of Country on either side. As with many Aboriginal groups, there are multiple sub-groups and dialects within the wider Barkandji language group.

PLANTS

FIGURE 4.1: Photographed on Barkandji Country c. 1879, this image shows (from right to left) Mary, Jacob, and their daughter, Doughboy, in a winter camp. It is full of insight if you know how and where to look.

This photograph speaks to me of the many aspects of plants and our interactions with them, as well as our exceptionally well-honed skills in using plants for vital and diverse purposes. It leaves me in awe of my mob, my ancestors; in awe of their ingenuity, but mostly of their deep knowledge of Country, evidenced in every aspect of their interactions with it. In this chapter I am going to take you through several aspects of the photo, which I will use as a tool to illuminate how central plants have been to every aspect of Aboriginal life.

This photo can be found in the Mitchell Library in Sydney and is part of a series of photos taken by Frederic Bonney, who came from and eventually returned to Staffordshire in England. Together with his brother Edward, Bonney annexed Momba Station, a western New South Wales pastoral sheep station, where the photo was taken. The brothers worked on and around Momba between the years 1865 and 1881, at a time when the pastoral industry had very recently encroached on Barkandji lands. The picture was taken in 1879 when Bonney was most active in his documentation. He had a deep respect for and friendship with our people. Unlike many who photographed and documented around this time, he did so with admiration – naming the people in the photos and writing about them with reverence and kindness:

> I wish to record what I have learnt of them during their better days, and hope that others, who have had like opportunities will do the same, so that sufficient information may be brought forward to prove their race to be better, nobler, and more intellectual than it is generally believed to be by those who have not lived among any of the tribes. All who have done so, and taken the trouble to learn something of their language, so as to better understand them, must have formed a good opinion of them.[3]

While Bonney's writing still evidences some problematic framing, he insists on the humanity of Aboriginal people, setting him apart from the majority of his era. I am so thankful this moment in time was captured, as I know many other Barkandji mob are also grateful. While as Aboriginal peoples our oral histories and our systems of passing knowledge through generations remain a foundational part of our cultures, documentation such as this can empower efforts to strengthen and revitalise knowledge of our people and of Country.

I first saw these photos in a wonderful book by Jeannette Hope and Robert Lindsay called *The People of the Paroo River: Frederic Bonney's Photographs*.[4] Robert Lindsay is known to my family, having lived in Menindee for many years, teaching at the Menindee School where I am proud to say my cousin Fiona Kelly is the principal. I met Robert many years ago and he generously told me some stories of my family that I did not know. As I was then very new to research, he also taught me how to surf Trove, pointing out many entries that featured stories related to the lives of my relatives. Robert has not only worked as a teacher for many years but has also undertaken extensive linguistic and historical work in both Menindee and Wilcannia, producing some important resources for the Aboriginal community.

Unlike many of the pictures of the time, Bonney's photographs are not overtly staged, and while we cannot know for sure, they seem to show people as they were, going about their daily business, not performing for a white audience. Photographic technology at the time required the subjects to stay very still for quite some time, so there is an element of staging perhaps to attain the stillness required, but it appears minimal throughout Bonney's series of photographs.

I do not claim that any of the people in the images are my known relatives, but they are my mob and I am fascinated to see them. Interestingly, I found an entry in Trove that ties my family to Momba some years later, in the early 1900s, the station having been given as an address to reach my great-grandfather in a newspaper advertisement. I have more work to do in researching my ancestors, but I have cause to wonder whether it was a longstanding connection that brought my family to this particular place.

When I look at this photo, I catch a glimpse of how my people were living at the time. I see and feel my ancestors looking back at me. I witness them here in this snapshot of a particular time, incorporating some new accoutrements (as can be seen, for example, in the use of the government blanket around Doughboy's shoulders, the canvas covering their shelter, and their European clothing and clay tobacco pipes) but continuing as they always have in many ways also. Quickly adapting to survive and, importantly, to stay on their ancestral lands by providing various services for the invaders on the newly established stations in return for food, clothing, tobacco and some semblance of safety from the violence of the frontier.[5] I see them here, innovating and surviving, but still maintaining a powerful connection to Country, despite the seismic ruptures of invasion.

Have a look now at the photo, what do you see?

This photo also illustrates the architecture of perception – that what we are able to see is heavily influenced by what we have been taught to see. There are some things we are unable to recognise, even when they sit in plain view. This is especially true of non-Aboriginal perceptions of culture, innovation, complex knowledge and our cultural continuance. Some may look at the photo and see people barely scratching out an existence. I look and see the abundance that comes with knowing your Country intimately: plenty of good food, the tools needed to procure, process and store it, and an ongoing connection to ancestral homelands and cultural practices.

I see many things within the picture, and invited my dear friend, the artist, poet, educator and fellow Barkandji wiimpatja David Doyle to also study and comment. What I see and what David sees is very much tied to the differing ways we have learnt and been taught. Broadly speaking, David was able to look much deeper than I could. I have not grown up on my Country; I have learnt more from books than I have from my own Barkandji elders. David, however, has much richer cultural knowledge because he has grown up on Country. He has been taught *on Country* and *by Country*, guided by elders and community members, the same way our Old People have learnt over thousands of generations. We often work in reciprocity with each other, combining our knowledge – I send him resources and books I come across to support him in his projects, and he will often look at work I am doing and offer a deeper cultural perspective. David is testament to the power of our traditional ways of doing, seeing and being, and his knowledge is far superior to mine, and always will be. I do not tell you this to run myself down or to say I am lesser as an Aboriginal person, but rather to highlight the foundational importance of access and rights to Country, for elders and young people to have opportunities to walk Country together, to fulfil custodial obligations, to learn from Country, and to strengthen and transfer knowledge intergenerationally. David speaks some language and has made a concerted effort to know and to practise his culture, also working as an educator alongside his art practice to keep knowledge and culture strong within our community, especially our young people. In many ways, our Barkandji community relationships embody altered forms of doing what our people have always done – coming together to exchange and share knowledge, to fill gaps, to teach each other; honouring and animating Country and our Old People together, still.[6]

GRINDSTONES, GRASSES, SEEDS AND BREAD

The series of photos by Bonney feature many individuals, but in this particular photo his annotations tell us we can see Mary to the right, in the middle is her husband Jacob, and to the left is their daughter, Doughboy. The first thing I see in this photo is food. In the hearth in the foreground there is an emu leg (from the knee down only) that has been left over from an emu cooked for a feed (Figure 4.2). Just in front of the fire area, with the foot of the emu seemingly pointing to it, David noticed a depression in the ground that looks

FIGURE 4.2: Detail highlighting the hearth clearing, emu leg and possibly a camp oven.

as though it may be a camp oven in the earth, with some oil staining around it, likely from cooked meat. Earth ovens are dug into the ground and use hot coals or heated clay balls to concentrate heat to cook. Traditionally, they were then covered with grasses, leaves or herbs, sometimes dampened with water to steam, before adding the meat, which was then covered with various plants to protect and sometimes to flavour the meat, or more wet grasses if steaming. Then more heated clay balls and embers were added before being covered with a sheet of bark and insulated using earth.[7]

The family also has a billycan, and Mary is perhaps having a cuppa as she prepares food (more on this on pages 90–3). If we look to the right, at Mary's hands, we can see that she is making something and that she is using what appears to be a sandstone grindstone to do so (Figure 4.3). While it is hard to know, Mary may have first scooped some seed out from her kangaroo-skin storage container using the coolamon (the oval-shaped, bowl-like container to Doughboy's left), then ground the seed on her grindstone (using the top stone in her hand) to make a paste or some flour, perhaps slopping water from her nearby cup into the mix. As Lesley's

FIGURE 4.3: Detail highlighting the grindstone and kangaroo-skin containers for consumables like water and seeds.

illumination in this book tells us, grindstones are an extremely important technology discovered and developed by Australia's First Peoples, allowing diverse groups to exploit and greatly benefit from their available resources. The development of grindstones has allowed us to make nutrient-rich pastes, and breads or dampers, and we know from various dating methods that this technology has been a part of the lives of the Indigenous peoples of Australia for tens of thousands of years.[8] Indeed, in a report published in 2020, archaeological work undertaken at the Madjedbebe site on Mirarr Country in Arnhem Land, conducted in partnership with traditional owners, has provided scientific evidence of the use of grindstones in processing plant foods going back 65,000 years.[9]

A large variety of plant species have been processed by our mob using grindstones. Native millet, sometimes called *Panicum* (*P. decompositum*), is a grain that was abundant on my Country and often used to make flour. From oral histories and the accounts of early visitors to our Country, such as Thomas Mitchell, Charles Sturt and others, we know that it was harvested for seed in large quantities. Native millet was stacked and subsequently burnt, with the seeds then easily collected from the ground. Purslane, sometimes known as pigweed (*Portulaca oleracea*), was also used, stacked in heaps and moved a few days later to reveal piles of seed, that, like native millet, was then ground and made into a dough to cook on hot coals.[10] Along with purslane and native millet, the seeds of various wattle (*Acacia* spp.) were often utilised, including one we call *malka* (*Acacia aneura*; mulga is the common English name) that has seeds that taste like sesame. The prickly wattle was also processed using grindstones, with a drop skin blanket used to collect seeds hit from their branches with a stick. Nardoo (Barkandji name *ngartu*) seed cases were also processed through being milled, ground and mixed with water to form a dough, then cooked. As nardoo is a food without high nutritional value, mostly eaten only during lean times,[11] it is unlikely this is what Mary has used to make her dough as we can see an emu leg in the fire, and the availability of animal meat generally denotes abundance as opposed to scarcity of food sources.

These seeds were stored by our mob also, and it is known that over winter in particular, the seeds of acacia, saltbushes, flax plants and grasses were relied upon. Archaeological evidence provides proof that seeds have been an important part of the subsistence patterns of people of this area, now known as the Darling Basin, for at least the past 15,000 years.[12] Frederic Bonney, who took this photo, noted that skin bags made from wallaby or a small kangaroo were used to store water and surplus seed harvests.[13] To the right of Mary's hand, which is resting on what may be a top grinder, perhaps grinding seeds to a paste on the larger grinding stone, you can see a water carrier made from kangaroo that maintains the shape of the animal. To the right of this water carrier, although hard to make out, is what David and I believe to be a kangaroo skin used to store seed. Our Barkandji Country is semi-arid, and to outsiders it may look like a desert, but it is extremely biodiverse. Since European record-keeping began on our Country in around 1880, droughts of twelve months' duration or more have been recorded around every six years, and the archaeological record also provides evidence of dramatic fluctuations.[14] The regularity of drought conditions attests to the skills of our people, their ability to thrive over many millennia in an environment where the availability of resources regularly and dramatically expands and contracts. Effective modes of storage for seeds, water and other food have been essential in our capacity to survive and to thrive.

As important markers of our history of skill and innovation, grindstones are another aspect of our culture that needs further illumination, as their usage opens up many opportunities for deeper understanding of the complexity and technical skill associated with our knowledge of Country. Dr Jonathan Jones, a Wiradjuri/Kamilaroi researcher and artist with whom I have been fortunate to collaborate on some recent projects, explains the centrality and importance of grindstones in a recent soundscape artwork he made for the *Emu Sky* exhibition, called *úntitled walam-wunga.galang (grindstones)*.

> *walam-wunga.galang* is a collaborative work with Uncle Stan Grant.[15] It's a work that celebrates a south-east cultural practice of collecting seeds, grinding them down and making bread. This is a practice that's been happening for countless generations in this region. In fact, in central New South Wales a grindstone was found at 32,000 years old, making us some of the world's oldest breadmakers. But like most Aboriginal stories, that's not part of Australia's history. So, in many ways, this project is about bringing those stories to light. Making these oversized grindstones to celebrate these big stories – these stories that are about our history.
>
> The works themselves are made from sandstone from the south-east that have been slowly ground down. And that process

of grinding stone, of shaping stone with stone, is about that enduring presence that we have. That slowly moving stone over stone, that connection we have that goes back for eons...

The work is thinking about a statement that Uncle Bruce has worked a lot with in his book, about Captain Sturt who was lost out at Coopers Creek. That he comes across a camp of Blackfellas who end up saving him and his men. And at night as he's been fed with roasted duck and cake, he sits in a house and he listens to the women of the camps grinding seeds, and he says that the sound they make is like a loom factory. And so, in that moment, we get a sense of how important these objects were in this region. How people were using these grindstones to feed their families, to feed our nations. And yet those stones have gone quiet. So in so many ways this project is about waking those stones up to tell these stories again, to feed not only our bellies but our imagination.[16]

THE COOLAMON

David also noticed that the coolamon[17] (seen to the left of Doughboy and in front of the other visible wooden tools) appears to have something in it, perhaps some recently collected plant materials. The coolamon is a container made from carefully chosen sheets of bark removed from both hard- and softwood trees. While also used by men, coolamons are most often used by and associated with Aboriginal women as a multi-purpose tool, used as carriers for food, small utensils and babies, and for preparing and serving food. All over Country, markings in trees can be seen where our people have arrived at places to collect food and extracted a coolamon from a suitable tree to do so. In areas where the seeds of trees and grasses feature heavily as staple foods, such as on Barkandji Country, coolamons are also sometimes utilised in the winnowing process to remove husks and chaff. Variable in size and shape, they may resemble scoops, suited to uses such as digging, or range to much larger trough-like bowls, suitable for holding significant quantities of food.[18] Many tools may be used in their manufacture, including fire to harden the wood; adzes, axes and rasps to cut and to shape; animal fats to seal; and heated irons to decorate. As women across diverse Aboriginal communities were responsible for gathering the majority of the foods eaten in families and communities, the coolamon was an essential part of any woman's toolkit.[19]

TRAPS AND NETS

To the right of Mary, resting on the roof of the shelter, is a large folded net (Figure 4.4). David and I both think that it is most likely this net is made from cumbungi (*Typha domingensis*), a plant that Lesley has written extensively about in Chapter 5. Cumbungi (sometimes also referred to as *Typha* or bulrush) is a very important plant on our Country, and in many parts of Australia, especially across the south-east. It is utilised as food and for technologies, which I write about in more detail shortly. Spiny flat-sedge (*Cyperus gymnocaulos*) and native hollyhock (*Malva preissiana*) were also sometimes processed for use in making nets, the latter particularly favoured for emu nets due to its superior strength.[20] David believes this looks like a net used to trap birds, particularly ducks, and he even spotted a bird feather stuck in it: although difficult to spot in the reproduction, it is to the right of Mary's head, trapped in the edge of the net, just hanging there. German zoologist William Blandowski, who also features in Chapter 2, made some important and rare observations and sketches, offering a precious glimpse into how large nets were tied to two trees on opposite banks of a river and used to capture ducks, sometimes 50–100 at a time.[21]

Throughout this chapter I have referenced a 1974 paper by archaeologist Professor Harry Allen, 'The Bagundji of the Darling Basin: Cereal Gatherers in an Uncertain Environment'. Although this paper is, in academic terms, somewhat dated, it remains relevant because it brings together multiple detailed accounts from the early visitors to Barkandji Country and the archaeological record to build a picture of our management and interactions with Country. Ethnobotanist Dr Beth Gott also used this method with great success, but with the important addition of incorporating the living knowledge that exists within Aboriginal communities.[22] Allen's method is also similar to the one Uncle Bruce Pascoe used to explore the narratives presented in *Dark Emu*. It is

FIGURE 4.4: Detail highlighting the large folded net, the scoop-style fishing net behind it and a branch of medicinal emubush (*Eremophila* spp.).

important to note here that there is a wealth of information held within living communities that is just as important as historical accounts, if not more so. I cannot help but wonder how much richer Allen's already impressive historical and archaeological investigation could have been had he also applied the living knowledge held within Barkandji community in his study.

Of the use of nets, Allen writes:

> During spring and summer, or whenever there was a fresh in the river, the Aboriginal population of the Darling Basin congregated along the river banks and lagoons. Specialized techniques involving the use of nets and traps were used to exploit the riverine environment. Fishing nets up to 90 m. in length and a metre wide, with a 75 mm mesh, have been recorded as being used either as seines with floats and weights or as fixed nets attached to stakes (Beveridge 1883: 44–6; Sturt 1833, vol. 1: 92) … Nets between 45 and 90 m. long and 18 m. deep were strung across watercourses to catch ducks and other aquatic birds (Krefft 1866: 368–9); poles were set up where trees were not available (Sturt 1849, vol. 2: 140) … Women generally collected shellfish, crayfish and reptiles and the men speared fish, but some other activities … involved the entire community. These included net fishing, the collection of bulrush roots (Mitchell 1839, vol. 2: 61) and the collection of grass seeds (Howitt MS 1862: April 17–21).[23]

The 'bulrush' roots referred to above are the tubers of the riparian[24] cumbungi plant (*Typha domingensis*). Growing in abundance along the edges of rivers and lakes, cumbungi is one of the most important plants used for food across the Murray-Darling river system. Our people ate the fresh flower shoots of cumbungi as a salad, but also dug out and collected the starchy root, steamed it in earthen ovens, then peeled and ate it. It is very tasty and has been likened to a type of potato.[25] When eaten, a fibrous residue is left, which is twisted into a knot and later chewed to work into a useful fibre. This remnant fibre, after being worked in the mouth was, according to Mitchell's account, dried on the roof of dwellings.[26] Once properly dry, it was then soaked and scraped clean using a shell, twisted into string and finally woven to make tools, such as nets and bags, and body adornments.[27]

On a recent trip to Country, Barkandji elder Uncle Badger Bates told me that the cumbungi tuber is very difficult to dig out. He explained that it is best to wait until the ground around it cracks open, making it easier to dislodge. Uncle Badger also told me that ancestral remains studied by archaeologists, some of which he has seen as part of his work as a cultural advisor over many years, have shown that the women in our Country chewed and processed the fibres of cumbungi so consistently that the teeth in certain parts of their mouths were visibly worn down as a result.[28] It is amazing to think that after thousands of years, these ancestors still hold visible clues evidencing this important subsistence and cultural practice, the deeply symbiotic relationship they shared with this plant inscribed in their bodies.

To process the fibre then twist it into string and weave it into large nets would have been a time-consuming practice. Here again the ingenuity of our people is evidenced – there would be no use in making such large and complex technologies if they would easily disintegrate. Our people, through careful observation and most likely a lot of testing, trial and error, deduced that cumbungi was the best fibre to use for nets because it is very strong and does not easily rot.[29] Originally coming from water, it is perfectly predisposed to be used in water without perishing. It is no accident that such large, labour-intensive nets were built from a fibre that was durable and expertly processed to ensure longevity and suitability for the important task of food procurement.

The next technology made from plants was one I did not see but that David pointed out to me. If you look to the top of the net there is another item, somewhat rounded, that is peeking out from underneath it. David has identified this as a scoop-style fishing net, noting that it has square knots (also noting the use of the different diamond-shaped knots in the technique used for the bird net). This second net has what looks like a bent opening made of a cane-like material at the top.

Technologies developed and perfected to utilise all available resources on our Country, which is often a place of harsh extremes (such as semi-regular drought), have been integral to our ability to thrive for the longest time imaginable. The foundational importance of plants in every aspect of our lives is evidenced in the fact that technologies made from plants even feature in important Creation stories, as told here by Uncle Badger:

Native Cat site, it's here at Tartna Point. This where the Moon site is, where they put the little boy in the Moon in the Paakantyi story, and also it's where the Native Cat used to breathe fire like a dragon and it singed all the beard off the women when they were trying to kill it because it was shaking all the fish out of the net, the fish trap, the women that hung around and kept looking down. The women with the beard, they walked away and the ones that wouldn't walk away, he burnt all the beards off them. That's why we always say the women with the beards, that's the ones, the Paakantyi women, we respect more than the ones without the beard, because they more knowledgeable.[30]

WEAPONS – CLUBS, BOOMERANGS, SPEARS

The weapons seen on the left of the picture are also all expertly made from plants. Next to Doughboy are two clubs that David, an accomplished carver, has noted are most likely made of a dark, heavy wood called *niilya* in language, or nelia (*Acacia loderi*) in English. This wood is very tough, a necessary characteristic of a club, and the weapon would have been carved from one piece of wood to maximise strength and durability. Nelia is also often used in the manufacture of digging sticks. Woods chosen are very carefully selected for their amenity for the task for which the weapon or tool is made. This knowledge has been carefully handed down over time. David explains how the clubs may have been carved at this time:

I would say that as this is after colonisation these weapons would have been carved with a steel axe traded from the Europeans. Traditionally a stone axe would have been used, with another stone used like a file to get the wood to the final shape before finally scraping with a sharp stone. After colonisation this final scraping with stone was often done using glass, working like a fine wood-plane with the wood coming off in strips and giving an amazingly smooth finish. The nelia wood has an incredible smell and grain, giving a beautiful shiny finish when scraped and oiled up. Uncle Badger and Anthony Hayward still scrape,[31] but use an old knife instead of stone or glass, although I have seen them use glass also.[32]

As with many other wooden artefacts, including digging sticks, these clubs would have been further strengthened through fire, which removes the moisture from the wood. Over time they would be cared for by rubbing in animal oils or fats to prevent splitting.[33] These clubs were often also decorated, reflecting the importance of mark-making and the inscription of cultural knowledge expressed through art. On the one closest to Doughboy we could just make out some longitudinal fluting close to the handle.

The two boomerangs next to the clubs are also used for hunting, and depending on their intended purpose are made from a range of differing woods. Like the word 'coolamon', the word 'boomerang' seems to be a type of Aboriginal English that has developed and is now commonly employed to describe this tool across many diverse communities and language groups. While some boomerangs are made to fly and return, others are manufactured to be used without throwing. Across Australia, boomerangs have long been used for a multitude of purposes, including as clap sticks and props for ceremony, or for fighting, hunting and play.[34]

Here, Uncle Badger explains the need to carefully consider the fundamental shape and grain of the wood to ensure its suitability for the intended carving:

> The roots on the river bank, they grow out and drop down, ones like the one there that grow straight out and then drop straight down. That's a good comeback boomerang, and it's pretty wood. The coolibah and the black box, they better, harder, than the river red gum root. They're the best, they've got the right bend, all the grain run the same way, and when you throw it, it don't break, but even a bent limb is good. But if you cut a big slab off and cut your boomerangs out, they'll break, because you're cutting across the grain. They'll crack across. But these ones with the bend, the elbows, the grains going the right way, and you just hotten them in the hot ashes then, and they'll go hard.[35]

David pointed out that as the boomerangs seem to be made of wood that is dark in colour, without much light sap visible, it is likely to be nelia, the same wood used for the clubs. He suspected the boomerangs would likely have been decorated with engravings made using a sharp stone, mussel shell or glass, as per the stylistic techniques Barkandji carvers have used for millennia, and continue to today. He used a magnifying glass to study them for this stylistic flourish and discovered some beautiful 'hidden' cultural information. On the boomerang closest to Doughboy, two circles have been carved into the wood, which David suggested might be depictions of two culturally important lakes, swamps or billabongs. On the boomerang furthest away from Doughboy, David found an exciting decoration that ties this maker's work to the work of Barkandji carvers today – an engraving of the Ngatji, running from the top left corner to the bottom right. The Ngatji are the two Rainbow Serpents who created our Country, and who still reside in the rivers today. They are a common motif in all Barkandji art, part of our Creation story and revered as living beings who must be respected through our care of Country and the waterways they live in. The Ngatji seen here symbolise our cultural connection to each other and to Country over time, one that is not lost, since so much of our cultural world revolves around these Creator Beings.[36] Their presence here also illuminates the centrality of our cultural beliefs to all parts of our existence and to every interaction with Country. Our cultural items are made with knowledge and respect for our Law, which attests to the knowledge and stories of our Old People over time, signposting our foundational responsibility to and reciprocity with Country. Mark-making and all forms of artistic practice honour, animate and tell the story of Country as our living relative, our mother, whom we must look after, just as the Ngatji must be cared for through our continued active custodianship of Country.

The three spears seen to the left, two of which are barbed, may have been made using a wood commonly referred to as belah (*Casuarina pauper*), known as *kariku* in Barkandji language. Another commonly used wood in making spears is mulga (*Acacia aneura*, Barkandji language *malka*), which is also favoured in the manufacture of boomerangs and various digging sticks for its strength.[37] The barbed spears may have been used for fishing, and the non-barbed one perhaps for hunting emus in conjunction with the use of nets. An illustration based on sketches by William Blandowski shows similar spears being used in an emu hunt.[38]

SHELTER AND EUROPEAN INFLUENCES

David and I both observed the kangaroo-skin rug on the ground in front of Doughboy, as well as a kangaroo cloak around her mother's shoulders. We also spotted a fluffy rug, most likely made from possum, under Jacob's foot. The appreciation and incorporation of European goods can be seen in the use of the items identified earlier, notably the canvas over the top of the structure and the tin billycan and cup, as well as the clay pipes the women smoke, European clothing, and a striped government blanket. It is difficult to guess what the women are smoking – it may be commercial tobacco traded or given as rations, or perhaps one of the many wild tobaccoes found growing across Australia, which are often mixed with the ashes of other plants to enhance potency.[39]

Uncle Badger indicated that while it is hard to see what the 'bones' of the hut are, the walls and roof are made of a thatch of cut bushes that, given the location on Momba, are probably either hopbush (*Dodonaea attenuata*) or 'bunta bush' (*Eremophila sturtii*).[40]

MEDICINE

David's much more nuanced knowledge of Country again came to the fore when we were looking at the shelter itself, as he was able to identify that not all the plants that can be seen form part of the structure. To the right of Mary, close to the folded bird net, is a branch of a very important medicinal plant, most commonly known as emubush. Emubush (*Eremophila* spp., sometimes known as native fuchsia) is used widely across Australia as a medicine in many different cultural contexts. This plant is one of the most important medicinal plants on Barkandji Country; used for various stomach complaints and specifically as a laxative, its leaves are made into decoctions for sores, taken internally for colds, used to treat skin complaints and infections, and as an eyewash.[41] It may be that Jacob and Mary are imbibing this emubush as a medicinal tea from their billy and cup, although it could also be old man weed (*Centipeda cunninghamii*) that they are drinking, a highly medicinal plant found on our Country and taken as a tea or made into a lotion to cure multiple conditions, but also commonly drunk for general health. Or it could also very well be the medicinal river mint (*Mentha australis*) found on our Country, used for coughs, colds and stomach complaints.[42]

Barkandji community members have shared much cultural and plant knowledge through open access online booklets made with elders and community members in partnership with the Lower Murray-Darling Catchment Management Authority as part of The Cultural Biodiversity Strategy: Paakantyi (Barkindji) Project. One entry for emubush compiled by community members states:

The leaves of this red-flowered emubush are eaten as a medicine for stomach complaints. The leaves are used to 'smoke' people for ceremony and to cure certain sickness. The leaves are burnt and the smoke is used to smoke the houses/residences of deceased people.[43]

Frederic Bonney also observed the medicinal use of emubush among the Barkandji, noting that

> Headache is a common complaint, and to relieve this a native ties to his forehead a small bunch of heated boughs; the fuchsia bush (*gooyermurra*) being considered best for the purpose. The same remedy is generally used to relieve pain elsewhere; an attendant holds the bunch of boughs while warm to the suffering part, and heats it again when cold.[44]

Its importance is also evidenced in its use in initiation ceremonies for young men, with Bonney recording:

> When the necessary preparations are made men and women go from the general camp to see the youth smoked. He and one of his companions sit or stand on a heap of green boughs from the fuchsia bush (*gooyermurra*), under which there has been laid dry grass and sticks: this heap is called a *windoo*, their word for an oven. The two youths are wrapped round loosely with a rug, their heads only being uncovered. After the dry grass and sticks at the bottom of the heap are lighted, thick smoke rises through the green boughs and collects round their bodies beneath the rugs.

After they have been smoked in this way the rugs are raised over their heads, so as to envelope the whole of them, the smoking continues, the youths placing a finger in each nostril to save themselves from suffocation. After a little of this they are removed from the *windoo*.[45]

Another plant noticed by David that I mistook as part of the structure is a branch of eucalyptus that can be seen stored in the roof, directly above Jacob's head. Like emubush, eucalyptus is a medicinal plant widely used across many Aboriginal communities. Jacob, Mary or Doughboy may have picked this sprig of eucalypt and brought it back to their camp to use on a wound, or to throw on the fire to get rid of mosquitoes or make smoke for a ceremonial purpose.[46]

DIGGING STICK – HIDDEN STORIES

Poking out behind the boomerang seen furthest to the left, and standing vertically, is what appears to be a large digging stick, known on our Country as *karnka*. It is apt that it is half hidden, as this all-important tool, so emblematic of Aboriginal women and their role in society, has been under-explored in representations of Aboriginal peoples and their cultural lives over time. As discussed in Chapter 2, many of the cultural items represented in popular narratives and within museums and galleries have been weapons. Further, many of the mostly male anthropologists who have (often with a very narrow view) written about Aboriginal people and culture have only had access to the cultural lives of men, as per the gender-segregated nature of Aboriginal social systems. This circumstance has led to Aboriginal women's culture and lives being much less well known and recognised.

My own learning journey is a perfect example of this. Despite being someone who reads a lot and asks a lot of questions, I have only relatively recently come to the understand the centrality, amenity and spiritual power of the digging stick. Not having had the opportunity to learn their true significance, I was probably unable to recognise them properly, even when they may have been right in front of me all along. The architecture of my perception was built to obscure them through lack of recognition and understanding.

Digging sticks are most often used by and associated with Aboriginal women, who have historically been responsible for supplying up to 80 per cent of the foods necessary to sustain their families and communities.[47] Forming an essential part of a woman's toolkit, digging sticks are used to obtain plant foods such as tubers, as well as small game, reptiles and fish, and are effective for stripping bark from trees, digging earth ovens, preparing the foundations of dwellings, as a prop in ceremony and dance, and as a weapon. They can often be used in conjunction with wooden containers such as coolamons, with the stick serving to loosen the soil, and the coolamon used to scoop out large volumes of earth. Digging sticks are made from the saplings of any strong wood. They are expertly shaped so that one end will often resemble a shovel, while the other is carved to a point for use as a multi-purpose tool. After shaping, one or sometimes both ends are hardened by fire. On our Barkandji Country, Uncle Badger explained that near the river, river red gum

(*Eucalyptus camaldulensis*) or black box wood (*E. largiflorens*) were most often used for digging sticks, whereas away from the river, malka or niilya were often favoured (see Figure 2.1, p. 33).[48]

Digging sticks were not just a tool to procure food but also to help cultivate the soil. For example, in digging up tubers such as murnong, the digging stick would act to till and aerate the soil. The tilling allowed the ashes from cultural fire regimes to provide nutrients to the growing murnong, with aeration of the soil also important for soil and crop health.[49] Many Aboriginal women continue to use digging sticks to obtain traditional foods. Today these tools are most commonly crafted from metal objects, such as crowbars, as makers ingeniously adapt classic designs by incorporating new materials.

The significance of digging sticks and the crucial role of Aboriginal women in the procurement of food is evidenced by their presence in creation stories centred on women. Some Aboriginal groups also included digging sticks in burial rights, underscoring the importance and centrality of this tool.[50] The digging stick can be found in communities right across the continent, despite the marked diversity across hundreds of Aboriginal and/or Torres Strait Islander nations and the massive variations in the physical attributes of Country.

Used to nurture Country, to physically and spiritually connect with it and to secure its bounty, digging sticks are also symbols of the centrality of women in food procurement right across our lands. But digging sticks must also be understood as women's weapons too, used to defend themselves and others. If you found yourself on any of the diverse homelands of the Indigenous peoples of Australia 250 years ago, I would surmise that you would rarely encounter a woman without one of these tools in her hand, and many colonial paintings and sketches feature Aboriginal women holding their digging sticks.

———

Just as the digging stick, and the portal into the world of my ancestors it represents, is almost hidden, there is plenty more that we cannot see because of gaps in research and the archaeological record. So much of our plant usage does not survive as physical evidence. However, as archaeologist Dr Aunty Sarah Martin explains,[51] earth mounds (used as baked-clay, heat-retaining ovens) found right across south-eastern Australia are a brilliant source of evidence of the foods Aboriginal people have been eating over time, and have been largely overlooked:

> Mounds have the potential to contain a wide range of palaeoecological information that informs about past environments, and the dynamics of management and exploitation of these environments ... Future research should concentrate on determining a regional chronology and extracting the full range of information from the heat retainers, artefacts, pollen, charcoal, starch, lipids, sediments, fibres, bones, scales, and whatever else can be found in these unique cultural deposits.[52]

We have much work to do to uncover more information about our plant use over time. But as Michael and Lesley's fascinating work shows, it is a matter of piecing together many different sources of evidence, and the growth in inclusion of the living knowledge of Indigenous communities in this work is an exciting path forward.

As Bonney's photograph illustrates, plants have long played a crucial role in every aspect of the lives and culture of Aboriginal and Torres Strait Islander peoples. We have used plants as highly nutritious foods, for powerful medicines and in developing a huge range of technologies for hunting, trapping, gathering, storing, preserving and processing, for creating shelter and for transport in the form of bark canoes. Our ability to survive and to thrive over the longest time imaginable has been greatly dependent on our capacity to understand and develop the resources of Country, and no resource has been more crucial than plants.

Today, some may see the use of plants for food, objects, cultural markers and so much more as a novelty, but perhaps also as unsophisticated. They may find it difficult to immediately see the abundance pictured here. It is important to understand that much of what our modern societies use today was originally derived from plants. For example, the medicines that now come in plastic bottles or sealed foil are largely processed from plants, or are synthetic versions of the medicinal properties in plants. Plants remain a critical part of everyday life around the world, but the way we use them has changed. They have become 'hidden', synthesised, still foundational to the end product, but now unseen. We have become too concerned with improving things that do not need to be improved, with the strange idea that the less it looks like the thing it came from, the better it is. This wish for separation, for dominion, to be 'above nature', has changed our human activities and contributed to many of the environmental crises we increasingly face. We have normalised systems that produce excessive waste and pollution, to make things that Country was already freely offering us, in reciprocity. Palyku woman Jessyca Hutchens beautifully explores what she terms 'relational abundance':

> The settlers often did not, or deliberately refused, to see the abundance of life and meaning woven across Country. The great lie of the colonial mindset was that it was the settlers who brought the means and ingenuity to find and produce an abundance that was lacking. To tame a savage land. To create an overflowing cornucopia of plants, seeds, tools, carriers, artworks, and their jointly told stories, is a starkly different picturing of Country. These living materials are not simply a 'harvest' of resources but create narratives around the reciprocal care and knowledge that allowed Indigenous peoples to thrive within the richness of their Country ...[53]

As we begin to too slowly realise our Earth is incapable of supporting the foundational tenet of capitalism – continual growth, continual taking – we begin to look afresh, to look back, to better recognise and see the value of plants.[54] Lesley's chapters on cumbungi and spinifex show a recognition of the utility of plants that is becoming more pronounced as non-Indigenous science comes around to our way of knowing. Indigenous knowledge tells us that if we work in reciprocity, giving back as we take, Country can provide everything we need.

There are surely even more clues within this photo that speak to my people's profound knowledge of all aspects of Country. Taking a deep dive into this picture, this moment in time, was so enjoyable and enlightening for both me and David. It led to countless excited yarns together on the phone and via email, and, luckily for us, Aunty Sarah and Uncle Badger also jumped into the conversation to give their perspectives and generously share their knowledge with us.

This journey has been a perfect example of how bringing differing perspectives and sources together can create a richer whole. My fondness for this photograph has only grown through our prolonged study of it in detail. Nowadays, I can look at it and trick myself, seeing Mary, Jacob and Doughboy come alive. I feel them with us today, and I hope they like the way we have told a small part of their story, honouring their deep knowledge and connection, their strength and their survival in the face of such upheaval.

5

CUMBUNGI

A BAG OF STRING

Typha, cumbungi, the bulrush, is a widespread but scarcely noticed plant in Australian cities. It makes itself at home in the damp cracks of the urban landscape – in drains, creeks, and pondages along railway lines.[1] Few of the people whizzing past those cracks, wondering how they are going to survive without single-use plastic bags, would be aware of the fibrous wonderland beneath their wheels. Yet Indigenous people along the Murray River made string bags, and effective fishing and duck nets more than 50 metres long, from the fibres of *Typha* roots. After they extracted starch from the rhizome (or horizontal underground stem) for food, the

FIGURE 5.1: Schematic diagram of a *Typha* rhizome and the string-making process.

remaining fibres were stored in bags used as pillows until needed for string-making. As told by settler-colonist Peter Beveridge,[2] 'When required for string-making, the knots were soaked overnight in water, teased out and scraped with shells of the freshwater mussel until they were clean, after which they were stored in small, neat hanks ready to be made into two-ply string.' (See Figure 5.1.)

Accounts of cumbungi use are piled up in the ethnohistorical records of south-eastern and south-western Australia, documented by a succession of settler-colonists.

> The natives gather the roots and carry them on their heads in great bundles within a piece of net ... The root is taken in lengths of 8-10 in., they peel off the outer rind, lay it a little before the fire, then twist and loosen the fibres, when a quantity of gluten, exactly resembling wheaten flour, may be shaken out. This gluten they call balyan. (Thomas Mitchell, 1839, Lachlan River, NSW)

> The staff of their existence is the bulrush root which the women gather among reeds, it is to them what bread is to the European. (George Angas, 1847, lower Murray River, SA).

John Mathew, a Scottish minister and anthropologist, described very large earth ovens holding 'half a ton' of rhizomes, prepared for large gatherings of people:

> A hole of circular outline 3–4 ft deep and 15–20 ft across was made. The roots were placed in the centre on a pile of dry wood. On the surface were strewn layers of long grass and light sticks. The fire was lit and the excavated earth returned as a covering. The time for cooking might be several days. When done, water was poured on the oven to cool it. (John Mathew, 1899, Murrumbidgee River, NSW)

Cumbungi contributed to relatively high population densities along the Murray River floodplains, where the mounds in which it was cooked constitute much of the visible archaeological landscape today. Clusters of earthen mounds are reminders of the relationship between *Typha* and large Aboriginal settlements, spanning thousands of years. When Thomas Mitchell headed down the Murrumbidgee from the Lachlan junction, he described 'lofty mounds of burnt clay or ashes ... [in which] the balyan is prepared'. On the Murray near Swan Hill, floodplain mounds are full of clay-ball heat retainers:

> When the fire had died down, the clay balls were removed with two sticks used as tongs, and the ashes were swept out. The hole was lined with moistened grass on which the rhizomes were placed, more moistened grass was used as a cover, and the baked clay nodules placed on top. The whole was then covered with earth until the food was cooked. (Peter Beveridge, 1889)

On the other side of the continent, on the wetlands of what is now the Swan Coastal Plain around Perth, cakes or bread were made from the 'farinaceous mass' formed by pounding the starchy rhizome, providing a staple food for the Nyungar.[3] Now captured as the suburban lakes of inner Perth – Lake Monger, Herdsman Lake, Lake Claremont – these wetlands offered a range of reliable resources, the staples being carbohydrates from reed rhizomes. Particularly important in late autumn when the starch content was at its richest, *Typha* provided a carbohydrate bridge to the yams (*Dioscorea hastifolia*) that came on a few months later (see Chapter 6). It was called *yandiyut*, and its root *yanjidi* – its presence 'hidden in plain view',[4] to use historian Paul Irish's term for Sydney, in surviving place names Yangebup, Bendya Yangeetup and Yanchep.[5]

If the usefulness of this plant is now invisible to most Australians, so too is the labour of the many Indigenous women who managed these plants. The imposition of European property relations turned the women from providers for their families and communities into thieves.

In 1991, archaeologist Sylvia Hallam retold the story of Fanny Balbuk, first documented by Daisy Bates in *The Passing of the Aborigines*:

> to the end of her life [in 1907] she raged and stormed at the usurping of her beloved home ground. One of her favourite annoyances was to stand at the gates of Government House, reviling all who dwelt within, because the stone gates guarded by a sentry enclosed her grandmother's burial ground... Balbuk had been born on Huirison [Heirisson] Island... and from there a straight track led to the place where once she had gathered jilgies [small freshwater crayfish] and vegetable food with the women, in the swamp where Perth railway station now stands. Through fences and over them, Balbuk took the straight track to the end. When a house was built in the way, she broke its fence-palings with her digging stick and charged up the steps and through the rooms.[6]

Hallam let Fanny Balbuk and her digging stick stand for the resistance of all the women dispossessed of their lands and livelihoods, and erased from the pages of history. Another important fibre in this thread of retelling is the work of ethnobotanist Beth Gott, who assembled the accounts above. They are just part of her decades of work on Aboriginal plant use of south-eastern Australia. These accounts must be continuously woven into stories if they are to endure into the future. For all their gaps and biases, these understandings need to be embedded sufficiently in our cultural memory to persist.

My focus here is not only the women who cultivated and managed the plant, but the plant itself, its many identities – hidden and visible – and its shifting relationships with people. I am interested in its potential as a larrikin plant of the Anthropocene to forge new relationships based on knowledge from past and present examples.

STARCH AND STRUCTURE: PLANT BEINGS AND BEING HUMAN

The capacities of plants emerge out of their ability to 'eat the sun', as Oliver Morton puts it.[7] Photosynthesis produces complex sugars and other compounds, allowing plants to store starch and form cellulose cell walls. Starch, such as in the *Typha* rhizome, provides energy for growth, while cellulose becomes fibrous, giving structure and strength to stems and leaves. Whatever it is to be human, it requires plants and their capacities. We are made by plants in the sense that they have provided the atmosphere that we breathe, and provide much of the sustenance that we eat. Microscopic *Typha* starch grains on European grindstones provide evidence of its use as food there at least 30,000 years ago, and use of other starchy plants is much earlier.[8] Rapid growth in our ancestors' brain size around 800,000 years ago needed an abundant supply of glucose; those who worked out how to cook starchy plant foods had ready and efficient access to this energy and the evolutionary advantage it conferred.[9] So plants have had agency in the ways our bodies evolved, and continue to be fundamental to our daily bodily relations.

The difference between animals and plants, and the lower status of the latter, has been one of the defining characteristics of Western thought since Aristotle defined animals as those who move and plants as those who do not. To see plants as immobile is to have a very partial view of what constitutes mobility. On one hand most plants are sessile, fixed in one place. On the other, they have moved through time and space, spanning continents over scales of millions of years. Pollen and spores of terrestrial plants disperse widely away from the rooted parent organism, allowing future generations to overcome potentially restrictive or difficult local conditions. Although the movements of plant parts (fruit, pollen, seeds) are widely considered to be at the 'mercy' of other agents of dispersal (animals, wind, water), plants capitalise on the movements of others to their own advantage. Even as they are rooted in place, plants move towards potentially beneficial and away from potentially problematic encounters with light, gravity, structures or water.

COSMOPOLITAN

Typha has been moving around since at least the Palaeocene Epoch 65 million years ago, carried on the remnants of ancestral supercontinents. This makes it a plant of the world, almost cosmopolitan.

The first European to describe it in Australia was French naturalist and explorer François Péron, who recognised it on Tasmania's Maria Island in 1802, where canoes were made from the stems, similar to the bulrushes growing around Paris. It was familiar to other settler-colonists, even if there was much confusion over the species (Swedish botanist Carl Linnaeus having only just come up with binomial nomenclature in 1753). Australian examples are just some of the hundreds of uses of *Typha* recorded across the world and throughout history. It was said to be as important to the Bolivians as bamboo to the Chinese.[10]

While Linnaeus was pinning it down in what became the Western scientific taxonomic system as *T. angustifolia* L. (said to be from the Greek *typhe*, meaning a cat's tail, smoke, or a cloud),[11] people all across the world had many names for it. These names reflect imposed identities of usefulness: corn dog grass, water sausage, Cossack asparagus. Beth Gott's forensic searches of the ethnographic literature found 109 names for the plant or its parts in Australian languages, from Bruny Island in southern Tasmania to Kalumburu in the north of Western Australia. A version of the Murray-Darling Wiradjuri term has stuck as the common name: cumbungi.

There are now understood to be two native species of *Typha* in Australia: *T. domingensis* (narrow-leaved cumbungi), more common in inland areas, and *T. orientalis* (broad-leaved cumbungi), more common in coastal and tableland areas. There are many difficulties in differentiating between these two species, with absolute identification needing chromosome numbers.[12] The Northern Hemisphere species *T. latifolia* has become naturalised in some areas since European settlement. In other words, it has adapted to and reproduced successfully in its new environment; it behaves as if it belongs.

Species identification and analysis of *Typha*'s genetic relationships are said to be difficult because its forms are highly variable and it hybridises frequently.[13] Though much still remains invisible in these DNA analyses, we can only wonder how millennia of human relations have shaped the plant and its parts, fostering the care and growth of fatter rhizomes, longer spear stalks, stronger string – just as the plant shaped us humans, providing carbohydrate, protein (via the nets), clothing and shelter on the precarious evolutionary journey. It may have provided other things too – Chinese medicine celebrates the capacity of *Typha* to treat atherosclerosis, heal wounds and improve wellbeing.[14]

FROM STAPLE TO WEED

As the seasonally flooded systems of the Murray-Darling and other Australian rivers became more regulated, the extensive wetlands and large stands of *Typha* and common reeds (*Phragmites australis*) that had supported many thousands of people were much reduced. New wetlands – drainage ditches, irrigation dams and channels, farm dams, urban interstices – offered expanded habitat in places where *Typha* had been less abundant. In a more intensively managed agricultural environment, it became common to use wetlands as water storage, or as part of water delivery systems, with excess water being delivered to the intermittent swampy woodlands with open canopies of river red gum (*Eucalyptus camaldulensis*).[15] As many red gums drowned, *Typha* took aggressive advantage of these opportunities, advancing onto new sedimentary structures and forming monospecific stands to produce a thick interwoven mat of stems.

Classifications of spatial belonging now became part of the picture. From being a staple food, *Typha* is now considered a native weed in the irrigation areas of both southern and northern Australia. It is the most widespread waterweed in North Queensland, where it is said by Biosecurity Queensland (the Department of Agriculture and Fisheries body responsible for weed control) to interfere with water flow to form swamps, reduce water quality, provide breeding places for vermin and mosquitoes, and reduce stock access to water. A number of herbicides are recommended to be sprayed for its removal.[16]

In the Anthropocene world, wetlands created by humans for all sorts of reasons, mostly agricultural, have become important habitats for birds, fish and other animals, preserved under the Ramsar Convention. Western Australia's Lake Kununurra is one such example,[17] formed on the Ord River for the supply of irrigation water to the surrounding area. *Typha domingensis* is the dominant emergent species in the adjacent Lily Lagoon, where it has a role as the lesser of two evils. Nutrient and sediment runoff from the nearby town create the perfect conditions for *Typha* to invade the open water patches. However, at times it has also acted as a barrier to contain the more threatening *Salvinia molesta*, a free-floating aquatic fern and a declared weed of national significance.

Weeds, like humans, occupy ambiguous space in (Western) thinking about nature. Weed evangelist Diego Bonetto has long encouraged us to have more respect for weeds, 'the strongholds of nature, the untameables, the unruly, the ones actually fighting back – blow after blow, seed after seed'.[18] But in doing so we need to acknowledge the complexity of our relationships with plants – adversarial as well as caring.

PLANT OF THE FUTURE?

In my survivalist moments, I wonder whether cumbungi, the resilient survivor of colonisation and urbanisation, constitutes the archetypal Anthropocene companion – unpretty, opportunistic, strong and able to thrive in variable conditions. We will need new relationships with plants, relationships not straitjacketed by the categories of the past. The bundle of practices that traditionally accompanied *Typha* cultivation included burning in late winter, harvesting, pounding, cooking, eating, chewing, rolling string on the thigh and making nets. The Anthropocene equivalents could include all of these and more: damming polluted waterways to encourage its cleansing presence, or creating unpolluted dams – or backyard ponds – to farm *Typha* patches. It could provide emergency or even staple food sources.

Climate change projections indicate that we may only have another thirty seasons of irrigated agriculture in the Murray-Darling Basin. We need to both reframe and remake agricultural understandings for this new world. Our concepts of cultivation are tied too closely to progressivist hierarchies of change and decline. They are not nimble enough for the future, with its uncertainty and dynamism, the possibility of surprise and sudden thresholds.

If the Anthropocene is characterised by volatile, unpredictable and rapid change, what does that mean for the relative certainty and seasonal rhythm of agriculture? What kinds of subsistence will be possible under changeable conditions that we only associate with hunter-gatherer lifeways of boom, bust and flexibility (albeit the specifics will be warmer rather than colder)?

Could *Typha* be a famine food for such conditions? Nutritional analysis suggest it can produce up to 7000 kilograms of rhizomes per hectare, up to 25 per cent of which is carbohydrate.[19] The flour has carbohydrate and protein levels comparable with corn, rice and wheat. But *Typha* is not a food without cost. It takes a lot of work to harvest and process. The rhizome tastes best in late autumn and early winter, when it is richest in starch. Interpreting the full range of ethnographic sources, Gott argued that:

> *Typha* was available year-round if necessary, but was more nourishing and attractive at certain seasons, implying, of course, that the food supply was sufficiently ample to allow choice to be made.[20]

In the literature on phyto-remediation, or plants as environmental healers, *Typha* is considered a botanical canary in the coalmine for its capacity as an indicator species of pollution.[21] It is used in Australian constructed wetlands to help remove effluent from natural waterways. Nitrogen, phosphorus, potassium, calcium, magnesium and sodium get drawn up into the leaves. One trade-off here is that dense *Typha* wetlands can create extensive mosquito habitat, especially in the tropics. It can also absorb and accumulate heavy metals such as aluminium, iron, zinc and lead.[22] Its 'naturally colonising' character renders it a suitable candidate to immobilise copper, magnesium and zinc in the tailings dams of a gold and copper mine near Orange, New South Wales. It has a dual role, reducing flow rates and allowing metals to precipitate onto the sediment, and then taking up the leachate, which is precipitated out as an iron and magnesium plaque on its roots.[23]

In Germany, *Typha* has been tested extensively as a building material in the renovation of heritage buildings.[24] The structure of its leaf mass provides a combination of insulation and strength via fibre-reinforced supporting tissue filled with soft open-cell spongy tissue. This can be manufactured into a magnesite-bound *Typha* board with high strength and dynamic stability, and low thermal conductivity. These tests draw attention to the potential industrial scale of its enormous growth rate and yield; an annual production rate of 15–20 tonnes of dry matter per hectare (four to five times that of evergreen forest production). We can imagine the moors and valleys of Germany covered with *Typha* farms, fulfilling the country's total demand for insulation and wall construction. The scale of landscape transformation that this companion might require is considerable, however. Similarly, in Senegal, crushed *Typha* fibres have been mixed with clay to provide a well-insulated building material that will reduce energy consumption and carbon dioxide emissions in the building sector.[25]

If we take inspiration from Fanny Balbuk and her digging stick, it is not because we can go back to her ancestral lands, but because we need to remake our futures. *Typha* was here long before us, and will likely be here long after we are gone. It does not need humans, although it has cultivated some productive relationships with us.

Its opportunistic and edgy nature has enabled it to benefit from management by Fanny's ancestors, and from our capacity to trash our planet. Resilient larrikin plants such as *Typha* may be more useful companions on the Anthropocene journey than we can yet imagine.

6

YAMS

Yams are the starchy edible tubers of the herbaceous vine genus *Dioscorea*, but they are also much more.[1] In this chapter I explore yams as windows onto wider social landscapes and engagements with Country. The underground tuber is only part of the plant. To those with no knowledge they seem to grow in hiding in the savanna landscapes of northern Australia; underground, invisible, out of sight and mind. To those who know what to look for, their presence is marked by many signs; at the tiny scale by their small above-ground stalks, which die back when the yam is ready to harvest, and at the broad landscape scale by rocky outcrops visited, worked and storied for millennia. To see old yam holes is to know of people and yams past.

My geographer colleague Jenny Atchison and I observed these processes with senior Murrinh-patha woman Biddy Simon and senior Jaminjung woman Polly Wandanga over several years in the late 1980s and through the 1990s. 'Collecting yams requires people to be attentive, watchful and patient', as well as having a strong arm, we observed.[2] The tuber can be more than a metre underground, connected to the surface by the vine stem, which itself can climb many metres. During the northern Australian dry season, when the vine loses its leaves, finding the actual yams requires expertise in recognising the thin dry stem above the ground. Biddy and Polly took us to a string of yam patches in the Keep River area of the Northern Territory. Each place had a name, and some of these patches were located along a Dreaming track and incorporated into its associated stories. 'With one exception, yam habitat was found on rocky, well-drained slopes in patches of monsoon rainforest. The exception was on flat sandy country surrounded by eucalypt woodland.'[3] None of these rainforest patches were large; often only a few tens of metres square, sometimes a couple of hectares. (Monsoon or dry rainforest is quite different to wet tropical rainforest, such as the extensive areas in the Daintree of North Queensland.)

> For Biddy and Polly, distinguishing a 'yam patch', or area of yam habitat, was an obvious and straightforward task. Both women remembered places in terms of the quality of their yams, and remembered their location in relation to other known places and resources.[4]

Dioscorea transversa, known to Biddy and Polly as *dangar* (common names in English include 'parsnip yam' or 'long yam'), is the favourite yam, requiring minimal preparation before cooking and with a sweet taste. It is widespread throughout the wet–dry tropics, where its range extends from tropical North Queensland, through the Northern Territory and into northern areas of Western Australia, including the Kimberley region. It extends down through southern Queensland and into the coastal rainforests, eucalypt forests, gallery forests and woodlands of New South Wales.

Throughout the north, dangar often grows in association with *garok (Dioscorea bulbifera*, 'round', 'cheeky' or 'white' yam), which requires processing to get rid of the 'cheeky' taste. Garok is commonly found in coastal, near-coastal, and riparian rainforests and monsoonal vine thickets (also known as dry rainforests), often on disturbed and marginal sites. Although the tubers are quite different – long and round respectively – distinguishing between the stems is more difficult. Dangar has a smooth blackish stem, while the garok stem is grey and rougher. This is somewhat easier during the wet season, when the dangar stem is brown and striped and the garok stem is white.

Collection of yam tubers is a highly seasonal activity, reflecting the plant's annual process of storing energy in underground organs in the drier months. In many places, this seasonality is also reflected in Aboriginal references to yam season. In northern Australia the collection period for *D. transversa* is during the dry season, commonly March through to August. Yam stems, and the clues they provide to the whereabouts of the tubers, shrivel and become more difficult to find, the further into the dry season. The availability of *D. bulbifera* is more variable, although still concentrated in the dry season.

Once the plants are found, digging yams is a labour-intensive process as the tubers of both species may be located up to 1.5 metres underground. You need to follow the stalk below the surface, digging carefully around it without breaking its trail, as the yam gradually reveals itself. Both a large metal crowbar and small axe were used by Biddy and Polly to lever rocks and cut away roots. We described the results of this process:

> The most striking result of the collection process was the large deep holes created ... It is estimated that anything up to 300 kg of stone and dirt were removed to retrieve an individual yam. This included rolling large boulders and rocks down the hill. Most of the smaller rocks and tree roots removed were placed carefully to the side and behind the digger, creating a ring of rocks around the hole. All sites except one had evidence of previous yam holes, the amount of sediment or leaf litter infill varying with the time since they were last dug. Biddy discussed these old holes a number of times in terms of people having used them over long periods of time:
>
>> you look all them hollow, all them hole, that where people bin sittin down, down old time ... they bin sitting down and coming in from all around ... (22 June 1998).
>
> The process of yam digging is likely to have had an impact on the soil profile at these locations, promoting organic matter and water to collect in the holes where yams had been replanted.[5]

Replanting is a widely recorded practice. As we observed with Biddy and Polly, on every occasion where it could be retrieved, the top of the yam was broken off and put back into the recently dug hole so the plant could regrow. On some occasions a few handfuls of dirt were also thrown in on top; however, the hole was never fully filled in. Biddy commented:

> ... when we break it [the top part of the yam] we put 'im back, let 'im grow a couple of years and when he grow the seed, he grow some more ... (7 August 1997).[6]

Similar practices are widely recorded in the research literature, which documents Indigenous people employing a variety of different methods of replanting or replacing parts of the removed tubers. Either the top of the tuber left adhering to the vine was replanted elsewhere, or it was left in place with the tendril intact. The hole could be left open to fill with friable soil and leaf litter, or the yam part and soil replaced into the same or even a nearby hole. Replanting does not only happen at the digging site. On one occasion a garok vine had bulbils – small bulb-like structures which form new plants – present. Biddy collected these and took them back to the garden of her outstation home at Marralam, a few kilometres away, where a number of garok that she had planted before were growing beneath the larger trees.

The quality of yams is often proportional to the difficulty of extracting them. Quality is assessed through the shape, sweetness and stickiness of the tuber. At one site where yams were easily and efficiently dug, Polly commented that she did not like the yams because they were too long and skinny. She gave up digging for them quite quickly. At another site, however, a single, thicker yam was pursued for over three hours.[7]

We noted 'the apparently symbiotic nature of yamming and quarrying' in the Keep River area.[8] Like many quarry sites visited by Aboriginal people for thousands of years to provide the raw materials for stone tool manufacture, these can be quite small outcrops of rock – quite different to the large stone or gravel quarries we are familiar with today. Stone tools are usually divided by archaeologists into flaked (e.g. spear points) and ground (e.g. grindstones, axe heads), requiring different types of stone. One site, Milyoonga, rises from the low-lying coastal plain as a steeply sided rocky ridge forming several hills with outcropping, very fine-grained rock suitable for flaking artefacts. If yam diggers have already excavated and stockpiled large blocks and boulders, they have made the task easier for stone tool manufacturers! Together with archaeologist Richard Fullagar, we observed a number of ways that yam digging and quarrying had operated together in this area, providing evidence of very detailed landscape knowledge and use.

It is crucial to recognise not all yams are the same – they grow, and grow differently, in many different landscapes, soils and climates. The third endemic Australian species of yam, *Dioscorea hastifolia*, sometimes known as the 'native yam', is found along the southern coastal areas of Western Australia. Locations include the alluvial plains in extensive yam or 'warran' grounds discussed in 'Landscape Transformation' (p. 124).[9] In contrast to the two tropical yams,

D. hastifolia has become adapted to the shorter growing seasons and lower temperatures of its more temperate geography. It grows during the winter and becomes dormant in the summer. In Western Australia it was collected throughout the year, but more intensely during spring and autumn, when rain made it easier to dig the soil.

These specific examples from our research exemplify wider trends. Roots and tubers have played important roles in human evolution, providing starchy staples to early humans and continuing to provide sustenance to billions of us today. The antiquity of yam utilisation by Aboriginal people in Australia is unclear, and difficult to trace in the archaeological record due to preservation issues, but some archaeologists such as Sylvia Hallam have argued that yam utilisation was part of the repertoire of the original inhabitants.[10]

One innovative way that recent research is tracing the long history of yam use is in the rock art corpus of northern Australia. Archaeologists Peter Veth, Sven Ouzman and colleagues compiled data from 3750 rock art sites across the Kimberley, finding that plants are depicted in about 15 per cent of these.[11] This is a much bigger proportion than in other parts of the world with strong rock art traditions. Yams are also part of the earliest rock art, dating back to around 36,000 years. They are particularly abundant in depiction over the past 10,000 years, which includes the period since the sea reached its present level and climates became warmer and wetter comparable with present-day (or at least pre-Anthropocene) conditions. This latter period is interpreted as showing a close symbolic identification between people and plants – these carbs are not just about calories, but also show how yams are strongly tied to human identity. In some art motifs yams are depicted as human forms – or vice versa. Yam figures are also prominent in the rock art of Arnhem Land.[12]

Collecting yams enfolds these different temporalities – the unimaginably long ones of human evolution; seasonal changes of access, taste and abundance; and the processes of daily life into which yam digging is organised. The daily outstation lives of Biddy and Polly included a range of activities, such as travelling, hunting, fishing and tending to children and relatives, as well as gathering plants. The time and effort expended on collecting yams needs to be examined in the context of these other pursuits. Tending to small children, for example, is a major consideration in planning yam collection and processing duties. To the extent that women's gathering activities affect plant biology then, they do so through everyday practices in specific places.

YAM COLLECTION AND PROCESSING – THE WORK OF WOMEN

This food source highlights the role of women's labour and knowledge. Women have maintained the food supply through environmental knowledge, connection to Country, and memory, and in many contexts continue to do so. As we alluded to in the book's introduction, the invisibility of plants in human history is closely tied to the invisibility of women. Collecting yams was and continues to be almost universally undertaken by women, often accompanied by children. Although men were observed regularly digging yams in

Western Australia in much the same manner as women, most of the literature records only observations of women. Only anthropologist Jane Goodale's account of the Kulama ceremony on the Tiwi Islands describes a gender division of labour in detail.[13] Kulama yams were dug by men with ceremonial painted digging sticks and then elaborately processed, again by men, to use in an initiation ritual. For the rest of the year the women continued to dig the edible yams.

As described earlier, digging yams is invariably a labour-intensive process. Where the yield has been measured and weighed by researchers, the results are variable, ranging anywhere from 0.4–3 kilograms of *D. transversa* per hour, with extraction being especially difficult and laborious in rocky soil.[14] Not all searches yield results – women might also spend hours looking and digging only to retrieve very little or even nothing.

There are no time estimates available for collection of *D. hastifolia*, although colonist-explorer George Grey observed c. 1837–39 that digging for yams was a time-consuming activity.

> The labour in proportion to the amount obtained, is great. To get a yam about half an inch in circumference and a foot in length, they have to dig a hole above a foot square and two feet in depth; a considerable portion of the time of the women and children is, therefore, passed in this employment.[15]

Yam yield per unit time, and thus the calorific efficiency of yam digging, seems then to vary mostly with the rockiness of the ground. Further research could test this proposition more systematically, but for the moment high calorie return on labour invested should not be assumed. If this is the case, other reasons (e.g. sweet taste, status and identity of yam diggers, other connections to yam places, lack of other food choices late in the dry season) account for the favoured status of yams as food.

Dioscorea transversa is commonly eaten without any preparation, or after light roasting. As mentioned, most accounts of this species are that it is sweet and very pleasant tasting, juicy and highly sought after. *D. hastifolia* is similar. *Dioscorea bulbifera*, on the other hand, requires extensive processing to leach out the bitter compounds before it can be eaten. Records show there are three variations in processing techniques for this species. The first method involves a sequence including baking, mashing or grating (using snail or baler shells), straining, washing (usually in running water) and in some cases finally straining again. The second method omits the mashing and straining steps and instead the cooked, soaked and washed slices or chips are eaten. Variations on the leaching techniques involved placing the tuber gratings or pieces in netted or string bags in running water, or alternatively, laying out the cut slices onto mats, which were covered and then submerged in water. The third distinctive processing method, recorded from the Kimberley, involves the peeled slices of tubers being coated in wet gum or eucalypt ash and then being baked in a pre-prepared bed of coals and ash overnight.[16] The coating ashes become very bitter and are washed off before the slices can be eaten. There are no reports in the literature on the consequences if this processing is not done. In analysing the distribution of yam art in Arnhem Land, archaeologist Judith

Hammond notes a particular cultural significance of the round yam in the pictures. She suggests that

> these clustered motifs may relate to ceremonial activities including local initiation and increase ceremonies designed to convey power and strength or reflect the toxicity of Yam flesh.[17]

LANDSCAPE TRANSFORMATION

The activities described previously to ensure the regular availability of yams, and in some cases to increase supply, have been drawn into historical debates about the extent to which Indigenous people transformed their landscapes and undertook 'cultivation' or 'agriculture'.

Before we return to those debates, it is important to note that the scale and types of landscape transformation were very different between the temperate and tropical areas. In the south of Western Australia, records show that the collection of *D. hastifolia* yams took place on a very large scale. The 'warran' or yam grounds described by Grey in his 1837–39 journals were so extensive that they occupied large portions of the alluvial terraces along the streams on the coastal plain, and even impeded movement.[18]

> ... for three and a half consecutive miles we traversed a fertile piece of land, literally perforated with the holes the natives had made to dig this root; indeed we could with difficulty walk across

> it on that account, whilst this tract extended east and west as far as we could see.[19]

Grey commented that he had seen these grounds in conjunction with 'superior huts, well marked roads [and] deeply sunk wells, spoke of a large and comparatively speaking resident population.'[20] Sylvia Hallam reported that families in this area held harvesting and digging rights in their own particular patches of *D. hastifolia*.[21]

In tropical areas, *D. transversa* and *D. bulbifera* yams are found in much smaller patches. These can nevertheless represent active landscape transformation by yam diggers. One way to understand this is by what happens when people stop digging. One of the sites we visited with Biddy and Polly, Mumburrum, was of particular interest as the yams were missing. The women remembered that yams had existed there at some time in the past, and we could see old yam holes. Several factors appeared to have contributed to yam decline. One was the impact of cattle, which graze in all of these areas and trample stems. The other was changing patterns of collection by Aboriginal people; regular use of yam patches appears to contribute to yam quality through the action of turning over the soil and rocks. Removal of these activities – such as when people are forcibly removed from Country – contributes to yam decline.

Another activity impacting yam availability is the management of fire associated with yam habitats. Sylvia Hallam's arguments are the most detailed in this regard.[22] Using a combination of historical and archaeological evidence, she argued that Aboriginal burning 'opened up' vegetation, leading to considerable landscape impact and the creation of new environments suitable for yam growth on the Swan coastal plain. Hallam also maintained that, in the context of wider landscape burning, particular areas where resources such as yams were located were carefully left unburnt. This combination of fire management in association with repeated digging of *D. hastifolia,* she concluded, extended and then intensified the range of *D. hastifolia* on the coastal plain. Hallam's argument that Aboriginal burning led to sedimentation of the floodplain nearly 40,000 years ago must be considered an untested hypothesis in the absence of further archaeological, geomorphological and palaeoecological testing.

In the north, a number of researchers have also reported fire being used to protect yams. The margins of fire-sensitive rainforests (also described as 'jungle patches') in the Northern Territory and North Queensland were burnt early in the year to protect carbohydrate-rich resources such as yams from hotter, more destructive late-season fires.[23] Other scientists, working in collaboration with Indigenous people, have reported that northern jungle patches where yams are located are protected by burning out from the margins.[24] These rainforest patches do not only contain yams, but a number of highly valued fruits, most of which appear in the early wet season. The common names in English of some of these fruits – black plum (*Vitex glabrata*), coffee fruit (*Grewia breviflora*), bush mango (*Buchanania obovata*), wild pear (*Persoonia falcata*) – indicate the richness of the seasonal resources that were being managed and protected. (The names also indicate something of the process by which early colonists engaged with the environment through their own frame of reference, often wiping out detailed Indigenous knowledge in the process.) Archaeobotanical evidence from a number of sites shows that rainforest patches provided consistent resources, and ecological refugia, through climate change on scales of thousands of years, as well as the seasonal changes. India Dilkes-Hall, Sue O'Connor and colleagues have shown a 47,000-year record of rainforest patch use at Carpenters Gap in the Kimberley.[25]

There are few descriptions of customary regulation of *Dioscorea* species. The most elaborate details of customary regulation of yams are recorded by Jane Goodale, in her study of the Kulama or yam ceremony from the Tiwi on Melville Island.[26] The ceremony was held at the end of the wet season and only after its conclusion could women and non-initiates eat yams. According to Goodale, the Tiwi believed that the correct performance of the ceremony would result in all kinds of yams growing plentifully. She inferred that the ceremony was highly significant to the maintenance of the elaborate detoxification knowledge for a species (*D. bulbifera*) probably only utilised during food shortages.

YAM LANDSCAPES

The practices associated with yams and yamming are variable and nuanced in different places and times, and with different species. It would be misleading then to generalise to the continental scale, and to the span of human history, without foundation. Two important themes thread through the previous sections. First, geographic variability in yam availability and habitat means that there is commensurate variability in the scale and nature of landscape transformation as a result of yam digging. Second, the role of landscape knowledge emerges as being extremely important to food supply and regulation. In using the term 'yam landscapes', Jenny and I were referring to the physical, social and conceptual dimensions of landscape.[27]

The smaller northern patches are not on the agricultural scale that at least some of the south-west 'warran' grounds seem to have been. The former provide a good example of the concept of storage in the landscape; as Tiwi women told Goodale, they did not need to plant yams back at camp because they always knew where some could be found. Such an approach is nevertheless dependent on the maintenance of memory, techniques and access. In the north at least, the strong seasonality of tuber availability, in combination with its patchy distribution and sometimes tenuous visibility, means that the traditional ecological knowledge surrounding collection and retrieval is very significant in maintaining yams as part of the food supply. The specific knowledge of how to process *D. bulbifera* and make it edible adds to this significance. The investment of effort to do this seems considerable and may be inefficient if understood solely in terms of calories. Yams should not be assumed to be easy calories just because they are starchy. Other factors such as the sweet taste and the connection to Country that digging facilitates may be just as important. Yam collection needs to be understood in terms of its gendered nature – the other activities and broader knowledge about the landscape, parallel activities and other rituals of stewardship, which were facilitated during women's collecting activities.

In planting *D. bulbifera* in an outstation garden, senior Marralam women were perhaps acknowledging, in contrast to the Tiwi in 1971, that by the 1990s it was no longer easy to find them in the wider landscape. The memory and connections they are attempting to maintain are not just to a food source but to their Country.

Active and regular use of northern Australian yam patches seems to have been important to maintaining a supply of good quality yams. Disruption of access in the post-contact period has led to perceived variability in the health of patches, with declining yam quality both in those that are never visited and those that are over-visited. Knowledge and skill in yam digging contributes to the status of many older Aboriginal women, but disruption of access and other social changes reduces the transmission of this knowledge to successive generations.

Given the threat of climate change to our agricultural systems, particularly vegetables, which have intensive water requirements,[28] the possibilities provided by yams and other tubers that are adapted to dry conditions are significant. Researchers have referred to sweet potatoes in this context,[29] but there is lots of potential to further consider yams. It may yet be that the knowledge and skill of Aboriginal women maintained and passed down over thousands of years helps to enhance food security for future generations of Australians.

7

SPINIFEX

Grasslands dominated by spinifex, an estimated sixty-nine species of the genus *Triodia*,[1] cover more than a quarter of the Australian continent, in the arid and semi-arid zones. Yet Western science has scant knowledge of *Triodia* physiology.[2] As the alternative name porcupine grass suggests, they are not comfortable to walk through and were cursed by early European surveyors. For pastoralists these grasslands provide fodder of last resort during drought times. Spinifex is a classic example of a plant with a rich history of human interaction and considerable future potential, yet one that is invisible to most Australians.

These extremely versatile plants have dozens of uses for Aboriginal people, including as an important fibre and adhesive. Endemic to Australia, spinifex are part of the widespread grass

(Poaceae) family. Other members of this family – bamboo, sugarcane, wheat and rice – are so entwined with human lives across the world that perhaps we might have expected settler Australia to consider more carefully the potential of spinifex to provide food and fibre. Fortunately, progress is being made to this end and archaeologists have begun examining residues on stone artefacts, some of them hidden for a century in museum collections.

They are among the toughest and most heat-resistant plants on Earth, yet fire is an important part of spinifex management, which Aboriginal people have been using for thousands of years. The leaf anatomy of spinifex allows it to be classified into twenty-seven 'soft', resinous species, mostly found in the northern and western parts of the arid zone, and forty-two 'hard', non-resinous species, mostly found in the south. Of particular interest is the resin of spinifex, which has long provided adhesive and other uses for Indigenous peoples. One notable application was for caulking wooden water containers, a technology vital to occupation of the arid zones. It is not surprising then that the resin became a valuable trade commodity. In an era of global heating, we all have much to learn about spinifex, and how we might use it in sustainable ways. This chapter discusses some of the ways that researchers have learnt about spinifex use, making clear its contemporary relevance.

USES OF SPINIFEX

Archaeologists Heidi Pitman and Lynley Wallis reviewed dozens of historical and ethnographic records to provide an overview of the diverse ways this versatile plant was used in the past, and continues to be used today:[3]

USES	ARTEFACTS
RESIN	
Hafting to handles	Spear heads, stone axes, adzes, knives, spearthrowers
Plugging/mending/caulking	Coolamons, shields, other containers, baskets
Attaching ornamental features	Including feathers, beads, shells
Attaching strings	Attached to pointing bones, bullroarers, shell pendants, other ritual objects
Making figurines and sculptures	Including way finders, ceremonial objects, and post-contact/contemporary art
Medicinal applications	Attached to sticks for warming and body application; as an inhalant, decoction, post-natal uses
Insect repellent	Burnt to repel mosquitoes
Trading in transportable formats	Large cakes and balls with stick handles
Toys	Balls for spinning game; toy dogs
LEAVES/FIBRES/FLOWER STALKS/RUNNERS	
Shelter cladding	Both singular and composite cladding systems
Protective cladding for resources	Piled over collected foods, e.g. seed pods, for protection
String manufacture	Nets, baskets, bags
Anti-spillage during water transport	Shoots placed in coolamon to minimise water loss during travel
Seeds as food	Seed cakes
Fish dams	Static and mobile dams

USES	ARTEFACTS
Way finding	Nocturnal torch, signal fires
Making fibre sculptures	Runners for baskets, leaves for 3D objects, including sculptures
Toys	Toy spears, sparklers

Pitman and Wallis noted that while varying levels of knowledge of use are retained in Indigenous communities, museum collections have become an important part of the historical evidence. Objects collected and stored in museums reveal a multitude of uses including ornamental, medicinal, structural and ceremonial functions. Two of the most evocative examples of museum specimens they discuss are an (undated) personal adornment object made from red *Erythrina* seeds and attached to human hair with spinifex resin. This object is currently in the South Australian Museum and is thought to be from the Dalhousie Station area, near the South Australia–Northern Territory border.[4] A second evocative object is a figure of a dog belonging to an ancestral being, probably crafted from spinifex resin. This was collected from the Killalpaninna Mission area in the Lake Eyre basin in the late 19th century, and is also in the South Australian Museum. Interestingly, art practice remains one of their common uses to this day.

Collecting and processing resin is complex. In his compilation from many sources, including his own extensive experience in the Kimberley and Western Desert, anthropologist Kim Akerman draws attention to the important role of collecting suitable plants.[5] The work involves locating appropriate spinifex tussocks, harvesting clumps of the grass and then taking them to a threshing floor, which could be a flat rock or an earth floor that has been cleared and prepared. The leaf epidermis of the twenty-seven soft species contains specialised resin-producing cells, with resin being exuded from the base of each leaf sheath. Pitman and Wallis distil the findings from many historical and ethnographic records to detail the process of extracting the resin and making it useful as an adhesive:

> The primary and most widely spread means of resin processing is via threshing of soft spinifex plants to remove the adhering resin in a dry, dust-like form, winnowing and yandying to remove extraneous sediment and chaff, and then gently heating the residual material until it coagulates into a viscous form that hardens upon cooling.[6]

If it is overheated, the resin develops a grainy texture and loses its effectiveness as an adhesive. We can imagine the level of experimentation in different environments that went into building this knowledge over long periods of time, and how it might have been exchanged and shared with other communities. Large-scale production of resin seems to have been mostly the work of women:

> While men may process small quantities of resin to fill immediate requirements, women, who are generally much more proficient at winnowing or yandying large quantities of material, are responsible for resin production when a greater mass of material is required for stockpiling or gift exchange.[7]

One of the most important uses of resin was hafting, modifying stone tools by adding a grip for more comfortable or efficient use, such as in an axe handle. Or attaching a handle for mechanical advantage, such as in stone-tipped spears. Archaeologists across the globe have been keen to trace hafting in the archaeological record because of the insights it provides into the development of human cognition processes. Hafting implies a level of social innovation and learning, and provides a window onto the many associated technologies – heating and mixing adhesives, fibre technology, stone tool technology and organic tools of wood and bone.

Seventy-eight plant species have been recorded as being used for hafting adhesives in Australia,[8] sometimes combined with animal products such as sinews or dried macropod faeces. *Triodia* was one of the most widely recorded. Other important resin-producing plants are *Xanthorrhoea* (the grass trees widespread through the south-east and south-west of the continent), which exude resin at the base of their trunks. Indeed, the name *Xanthorrhoea* means 'yellow flowing', referring to this resin, which was combined with charcoal, animal or human hair, and dried kangaroo dung as binders.[9]

It is also worth mentioning the medicinal uses;[10] for example, making a medicinal bodywash by boiling spinifex leaves in water. This was used for itchy skin and sores, and was also drunk to relieve cold and flu symptoms. The ash of burnt spinifex was rubbed on sores, and smoke from the burning plant used to get rid of mosquitoes. Smoke from the heating phase of resin production was inhaled for respiratory complaints.

The earliest evidence for spinifex use is found in the 47,000-year-old sediments at the base of the Carpenters Gap rockshelter, near Windjana Gorge in the Kimberley region of Western Australia, where fragile stem parts have been preserved as part of a rich archaeobotanical record studied by archaeologists India Dilkes-Hall, Sue O'Connor and Jane Balme.[11] The relative abundance of spinifex and other plant remains shows how the environment has changed over time, and how patterns of collection have responded to these changing environments. In an area currently dominated by monsoon rainforest habitat, spinifex provides part of the evidence of expansions and contractions of more arid-adapted grasslands over time. They increased in the lead-up to the Last Glacial Maximum (approximately 24,000–18,000 years ago), the coldest and most arid period since people first occupied Australia. The Puritjarra rockshelter, excavated by archaeologist Mike Smith in the spinifex and sand hill habitats of Central Australia, provides another body of evidence of how these habitats have changed over the past 40,000 years.[12] In contrast to Carpenters Gap, which is more heavily influenced by monsoon rainfall, increases in spinifex at Puritjarra are interpreted as evidence of higher rather than lower rainfall. This indicates the waxing and waning of the spinifex grasslands in different places, but also their broader continuity over time. At Puritjarra the grasslands expanded in the past 10,000 years, and remains interpreted as spinifex remnants peak around 5800 years ago.

ADHESIVES AND FIBRECRAFT

We're going to focus on two particular case studies of archaeological research into spinifex, adhesives and fibrecraft. In these examples different groups of archaeologists undertook experiments with contemporary spinifex to better interpret the stone tool record. We can also see the role of museum collections in preserving evidence – often hidden for long periods of time before someone comes along and asks a new question of it.

As well as studying macroscopic plant remains (remains that can be seen with the naked eye) from human occupation sites, archaeologists also use microscopes. (This is comparable to how palaeoecologists such as Michael examine pollen from sediments, as he discusses in the Bolin Bolin chapter.) Given that stone tools are the best preserved archaeological remains, there are exciting opportunities to understand plant use by looking at the patterns of use-wear on the tools, and microscopic residue of plants left behind. These residues include starch grains from tubers, nuts and the trunk pith of plants. This kind of analysis has already provided an important means of understanding the long history of seed grinding in Australia.[13]

Hafting adhesives
Belgian archaeologist Veerle Rots is an international authority on hafting, having examined examples throughout the European and African archaeological records. She worked with a team of Australian archaeologists to examine the evidence for hafting adhesives

in Australia.[14] The team undertook experiments with different methods of adhesive preparation, different hafting techniques and on different types of tools. They used 'ethnographic recipes' recorded in the historical and anthropological record, together with controlled experiments, to understand hafting arrangements and breakage patterns used in the past. Understanding adhesive 'recipes' and their chemistry is also important for interpreting archaeological residues preserved on the surfaces of stone tools for thousands of years. They compared spinifex (*Triodia*), grass tree (*Xanthorrhea*) and a third important resin-producing plant, tangled lechenaultia (*Lechenaultia divaricata*). Lechenaultia produces resin in its roots, which were dug up and scraped down with stone knives to release the resin, which was then heated in hot ashes.

The spinifex resin for the experiments was collected from various parts of northern Australia in six different forms. These were then mixed with different proportions of dried kangaroo or wallaby dung and beeswax to improve pliability and stickiness. The resin was fashioned into different types of tools and used or operated to test the quality of the hafting arrangement. For example,

> spears were launched with a wooden spear thrower by a single experienced shooter into a uniform target (a termite mound, ~1m high) … Success and failure were documented by recording whether the point directly hit the target, whether the hafting broke and the number of shots per point.[15]

We can reflect on the dispassionate language of scientific experimentation in these tests, and imagine how that success or failure might have played out in the world of a hunter hunting a kangaroo with that spear. Not only was there a lot of work in developing and manufacturing an effective spear shaft and stone tip, but if the resin that held them together did not work, the whole endeavour was in vain.

The archaeologists found that different modes of collection and processing of the spinifex resin produced variable results. Lumps collected from the environment after bushfires were more brittle and not as sticky as spinifex threshed from leaves. Some of their samples had been collected from ant beds, and these needed beeswax to be added to be pliable and sticky enough. Their main conclusion was that the creation of an effective haft depends on many factors: 'resin source, heating regimes/temperature controls, additives, mixing processes (grinding, crushing and pounding) and methods of application all play important roles'.[16] This is yet another area where high levels of skill, a lot of experimentation, and being attuned to local conditions were all involved in the accumulation of Indigenous knowledge.

Fibrecraft
In another set of experiments, archaeologists Ebbe Hayes, Richard Fullagar, Ken Mulvaney and Kate Connell examined the residues and use-wear patterns of spinifex on grinding stones.[17] This particular case study led to the broader conclusion that archaeologists might have been missing a whole lot of stone tool use related to fibrecraft, due to the common assumption that

most grindstones were used to process food such as seeds. Direct archaeological evidence of fibrecraft is relatively rare, as string and nets do not preserve in deposits as well as stone and bone do. But the manufacture of string using animal and plant fibres is such a fundamental technology that archaeologists know it must have been widespread throughout the history of human habitation of Australia. Indeed, they consider it to have likely developed even before the emergence of modern humans.

Let's see how Hayes and colleagues did it. The story hinges on a grinding stone (about 50 × 30 centimetres) that had sat in the Archaeology and Anthropology Museum at Cambridge University since 1926. Again, the ethnohistorical record was used in conjunction with experimental approaches, this time to the single grinding stone. In this way the archaeologists could connect poorly preserved plant remains to well-preserved stone tools, via the wear patterns and microscopic residues that remain on the stone. Records of stones used to extract and process plant fibres are not common, but our friend spinifex is one example that stands out in these records, partly because of its multiple uses. Pitman and Wallis indicated three lines of evidence for the use of stone to pound spinifex to extract fibre.[18] As well as accounts from elderly Aboriginal informants,[19] they drew attention to the importance of museum records and collections. For example, museum labels associated with spinifex string samples in the Western Australian Museum suggested that the spinifex had been beaten ready for making twine.

Only one known museum specimen is recorded as having been used for processing spinifex leaves for fibre. This is sample

CMAA 1926.591, a grindstone from Western Australia. Pitman and Wallis had said, 'Whether this object might have been involved in spinifex fibre production, or whether the collector misidentified its use and spinifex seeds were meant instead of leaves is unclear ...'[20] Hayes and colleagues were able to undertake experiments in processing whole clumps of spinifex, and then apply their experimental data to this grindstone, which is still housed at Cambridge. CMAA 1926.591 was 'collected' from the Southesk (previously South Esk) Tablelands in the Warburton region of Western Australia during the Michael Terry Expedition of October 1925.

> The label questions NSW as the Australian state of origin and overwrites with the correct provenance as Western Australia. The catalogue card also has a question mark against food preparation and describes its use for grinding spinifex leaves.
>
> The artifact is described as a large grinding stone, possibly made by people of the Boonara (probably referring to the general term for 'those who live by winnowing grass seed in ['pan: a] or wooden dishes), Bunara, Boonara, Waiangara' ... However, grinding *Triodia* spinifex leaves would be for extracting resinous hafting adhesive or fibres rather than for food. The grinding stone is an indurated sandstone with two large grinding grooves on the upper surface (Surface 1), which range in depth from 29 mm (Groove 2) and 32 mm (Groove 1) ... The lower surface of the grinding stone (Surface 2) has not been ground. The tool appears to have been cleaned prior to storage at the museum.[21]

In this extract we again see a group of archaeologists having to lay out the terms for their experimental data to be reliable: careful consideration of the conditions under which the grindstone has been stored, the accuracy or otherwise of its label, critical assessment of the cataloguer's description, and their own systematic description. What we don't see is the conditions under which it was collected, or the colonial context that led to so many Aboriginal artefacts lying in museums on the other side of the world.

———

Michael Terry was an English-born explorer and author who wrote a number of books about his travels through 'remote' Australia.[22] We can discern something of the context of collection of CMAA 1926.591 in his book *Through a Land of Promise*, in which Terry and his group were constantly engaged in the search for water, such as at the rock hole at the Southesk Tablelands described here:

> On a grassy flat before the hole there were remains of a large blacks' camp. Low brush windbreaks were grouped in three and fours. Never before have I seen so large a collection; there must have been accommodation for at least one hundred. This was specially interesting as proof of the importance of these rockholes; seldom do more than a score camp together, so in normal times this must be quite a concentration centre.
>
> We collected a most interesting grinding stone, with which the lubras make flour out of seeds from spinifex, nardoo, box or

pepper grass. By rubbing a small stone backwards and forwards across a sandstone slab they soon make a groove. Into this the seeds are fed, soon to become a fine powder for baking as a kind of scone. A hefty digging stick, used for unearthing yams, hardened by fire at the flat point, was another treasure. Bark twine, a couple of kylies[23] and an old shield completed the collection.[24]

The sense of everyday activities suddenly abandoned, and Terry's mention of an encounter that evening with 'two blacks' dogs',[25] and the possibility that they were being watched, suggests that these were not 'remains' but evidence of an active and vibrant settlement. We can only imagine the woman who returned later to find her own treasured grindstone gone, having become part of Terry's stolen 'treasures'.

The researchers extracted microscopic plant residues (starch grains, silica, charcoal, cellulose fibres and various plant cells) from the surface of the grinding stone and compared these with the residues they had taken from experimental pounding tools they used to soften spinifex fibres. They found residues on the museum grinding stone to be consistent with its written documentation, which said that the tool was used for pounding spinifex leaves. 'The ethnohistorical records combined with our specific tool-use experiments suggest considerable overlap in the wear and residue patterns on stone tools used for processing clumps of spinifex for fibre and seeds.'[26]

The interpretation was not completely clear-cut for two reasons. One was that the residues, notably the phytoliths and starch, were consistent with several uses – pounding for resin, processing for fibre and processing seeds. The other was that grindstones are known to have multiple uses and are not necessarily specialised to a particular use. But their interpretation did show how previous assumptions about the uses of grindstones were too narrow:

> we suggest that at least some of the grinding stones (both upper and lower) previously identified as seed grinders may well have been used to soften *Triodia* spinifex fibres to make string for nets, bags and other craft products.[27]
>
> We argue that Aboriginal exploitation of *Triodia* spinifex for fibre was probably more common than previously thought, and that key to its exploitation and archaeological identification are reassessment of grinding/pounding stones, including handstones, hatchet heads, mortars, lower grinding dishes and bedrock grinding patches. We suggest that previous identifications of spinifex processing to grind seeds for food may be an error.[28]

My musing on and consideration of the museum history of the grindstone is inspired by Heidi Pitman's evocative narrative of 'A Cake of Spinifex Resin'. In researching museum samples related to spinifex for her honours thesis, she came upon and became very connected to one in particular, A64065, held in the South Australian Museum.[29] As with the grindstone, the cake of resin holds a poignant lesson that museum specimens, albeit often collected under conditions that were extractive, still hold important information and evidence. Pitman considered part of her work to be to re-enliven the resin by telling its story:

> Creativity and imagination combined with my interpretative endeavour as an archaeologist provided a means to expose the vibrancy of this object's history of human-thing encounter … enriched by narrative, this cake of spinifex resin today continues to be charged with meaning.[30]

Pitman told the story from the point of view of the resin:

> In this manner I had become the form I am today, a large cake measuring approximately forty centimetres in length by sixteen centimetres in width. I weigh almost 1.5 kilograms. A speckling of grass particles is visible across my smooth, slightly lustrous surface. My subtle, slightly sweet smell has lessened over time, but I remain exceedingly tough. I represent many hours of labour by skilled and knowledgeable women, made not to be used straight away but initially to become a valuable trade commodity before eventually being heated and broken into smaller pieces and used for a multitude of purposes.
>
> Alas, I was never to live out my intended use. My makers traded me to a man in return for some tobacco. He was unlike any person I had ever seen before. His skin was pale, and he wore strange clothes. He rode on a large animal that broke up the sand plain as it moved across the landscape. I was carefully packed

amongst other items made by my maker's people and transported by horse to the South Australian Museum. Who was this man who carried me south, away from my home, into Adelaide? Why did he choose me? What did I have to offer him?[31]

I became a museum artefact to live out the rest of my life wrapped in paper and placed in a cardboard box, rarely to be seen by people, let alone to be heated and moulded again. This year I felt the warm touch of human companionship, even valued again as I was carefully removed from my box to be measured, weighed, analysed, and photographed. Why only now, one hundred years on, am I seeing the light of day? Who is it that holds me now?[32]

The colonial history and context of museums and their collections is a fraught one, as Pitman is aware. She notes:

The South Australian Museum's collection of Aboriginal Australian ethnographic material is considered to be the most comprehensive in the world. Its Aboriginal Cultures Gallery has been criticised for its colonial artefact-based approach, where architecture and layout was argued to communicate a sense of fossilised objects, 'locating "real" Aborigines in a spatially and temporally remote space (Hemming 2003, 64).'[33] In more recent times curators have endeavoured to provide a more inclusive representation and display of Aboriginal cultures.[34]

NEW USES OF SPINIFEX, SUSTAINABLE HARVESTING AND CLIMATE CHANGE

One of the things that thinking about museum specimens does is to fold different temporalities together – past and present are evoked in ways that are intertwined rather than separate. This also helps us think about future possibilities in the use of spinifex. As Alison Page and Paul Memmott have outlined in their book in this series, *Design*, these possibilities include spinifex as a building material, and as an innovative biomaterial with new applications, such as extra-thin condoms.

The new possibilities of spinifex as a building and insulation material stem from recognition of its long history of use in shelters (see pages 132–3). The collaboration between researchers at the University of Queensland and the Dugalunji Aboriginal Corporation brings non-Indigenous scientific enquiry and Indigenous knowledge together to explore new research and business opportunities. Paul Memmott and colleagues discuss the usefulness of Indigenous knowledge in the project, explaining that 'oral histories from Indigenous people were not only insightful of ecological processes but also presented sensitivity to the ways these processes can be harnessed wisely'.[35]

Wuthathi/Meriam woman and intellectual property law expert Terri Janke cautions that carefully negotiated partnerships are crucial in these contexts so that Indigenous rights are protected and respected.[36] She explains that 'the impact of sustained colonisation has meant that, as with the land, Indigenous people came to be seen as not owning their plant resources and plant knowledge.'[37] Biotechnology, she explains, can transform cultural knowledge into new 'discoveries' that in terms of Western science and law can be patented and owned. Australian copyright legislation is particularly unfit for purpose as it does not protect Indigenous knowledge that is orally handed down across the generations (see Chapter 9).

The new and broadened uses of spinifex are likely to become more significant; its capacity to withstand temperatures of 50 °C means that it is likely to be a very important part of arid-zone ecosystems into the heated future. As well as the intellectual property issues, there are many other concerns to be addressed if large-scale harvesting of spinifex is to be sustainable, and knowledge on this subject is limited.[38] Harvesting techniques need to minimise disturbance to soil, vegetation and other biota. The season is also an important consideration; 'harvesting in the late dry season could facilitate new growth in the coming wet season, while harvesting during/after the wet season and before seed-set might hinder regeneration.'[39] Indigenous expertise, especially in relation to burning strategies, will be vital to management and harvesting. It is crucial that we give due recognition, rights and compensation to such expertise, rather than repeat the 21st-century equivalent of Michael Terry stealing a grindstone.

8

QUANDONGS

ZENA CUMPSTON

Quandong (*Santalum acuminatum*) is a very important plant on my Country, as it is across many areas of Australia.[1] While quandong grows in some southern parts of the Northern Territory and Queensland, their natural distribution is across the southern regions (spanning from the south-east to the south-west) of mainland Australia.

Fruiting between August and December, quandongs are sometimes referred to as wild peach, desert peach or native peach.[2] Because they grow well in arid and semi-arid conditions, quandong trees are valued all the more, and they are well known as one of the first native foods to be accepted and widely used by the European newcomers.[3] Their appearance when in fruit makes me feel they are showing off, much like an exotic interloper – beautiful, richly coloured crimson fruit popping across a backdrop of muted, smoky, arid tones. Look at me whydontcha, with my red lippy.

When I was a teenager my mum discovered a quandong tree close to our house – the thirty-ninth she had lived in since marrying our grass-is-greener eternal-wanderer dad – deep in Adelaide suburbia. Her excitement was curious to me. She was almost crying with joy, and I wondered how good this fruit could possibly be to provoke such an exuberant, full-body reaction. She collected lots of the fruit and made a jam. Her ingrate children, including me, rejected it: too tart for our deeply Westernised palates.

Today I would do almost anything to try that jam again, properly understanding only now, thirty-plus years later, what it meant to Mum. That it signified her longing for home, cherished memories of childhood. Culture. Family. Ancestral connection. And now for me, thanks to this childhood memory, quandongs will always remind me of my mum.

I don't know as much about quandongs as my mum, but what I know I will happily share with you. First, quandongs really are exotic interlopers because they are parasitic – they can only grow if they are able to attach themselves to another host plant, most often acacias and saltbush. This renders them somewhat difficult to make commercially viable, but isn't that the point of being an exotic interloper anyway? One should always remain reassuringly expensive if one is to maintain such appeal, and being difficult to 'tame' certainly assures this.

In many places quandong trees appear to grow in groves, providing evidence that the Aboriginal people in a range of landscapes had learnt to successfully propagate and plant them to make them viable en masse.[4] Their deep knowledge of their Country and the quandong tree enabled them to make this important food source both available and abundant – indeed, grown in such large numbers that technologies to process them were necessarily invented, as we explore below.

My friend David Doyle packed some quandongs in his daughter's lunch box recently, which stirred a lot of interest from her teacher and classmates and allowed David to share some of his cultural plant knowledge, which was greatly appreciated. Like me, David also loves the quandong fruit, but unlike me he has access to it on Country. In one month at the end of 2021, he made quandong wine and quandong port, mead and syrup, as well as quandong jam, chutney, chilli sauce, salad dressing, ice cream, pavlova and pie, a quandong-and-samphire relish, and even sweet-and-sour-pork with quandong. Our Old People would be proud!

LESLEY HEAD AND ZENA CUMPSTON

Quandong is part of the sandalwood (Santalaceae) family, highly prized across the globe for its aromatic timber. The combination of fruit and nut provides a staple that can be processed and stored: the tart, tasty fruit can be dried; the oil-rich kernel inside the nut provides a liniment for muscle aches and pains, as well as a facial moisturiser; and the bark and leaves have microbial properties.[5]

Quandong fruits vary in size from 15–25 millimetres. The outer flesh is 3–4 millimetres and rich in antioxidants, vitamin C, vitamin E, magnesium, zinc, selenium and iron[6] – making them 'really good when you don't have meat', as Central Australian man Pompey Everard described.[7] There are many records of Aboriginal groups harvesting quandong fruits in large quantities and processing them for later use. Processing included drying, pounding and rolling into balls.

The single round nut inside the fruit has a very hard casing with an extremely valuable kernel inside. Central Australian woman Mollie Everard explained the medicinal uses:

> The edible part on the outside, we'd gather and eat. And as for the stones – we'd collect them too, hit them with a rock, prise them open, and get the kernels – you know, the inside part. We'd gather up a whole lot of them. My grandmother would gather them up and grind them. She'd grind them in a *wira* dish, after chopping them up well. It's to rub into aches and pains. You know, the kernel inside. The skin on the outside they'd eat as food. And they would rub the kernels into the back. If you had a backache, you'd rub it in. Sleep the night, then get up and think: 'Hey! That pain of mine has gone!' With the aid of that medicine, I've recovered. They'd just get better and go off for game.[8]

Medicinal and cosmetic uses of the kernel are also documented by archaeologist Colin Pardoe and colleagues in Barkandji Country:

> [it was used] to relieve tooth ache and gum boils (Beryl Carmichael of Menindee, personal communication). The crushed kernels were also used among Barkandji women of the Darling River as skin cream and hair conditioner (Dayle Doyle and Dot Stephens of Menindee, personal communication).[9]

These diverse uses have also been widely recorded in northern Victoria, as ethnobotanists Beth Gott and Nelly Zola explain in their book *Koorie Plants, Koorie People*:

> Aboriginal people ate both the fruit and the seed which sits inside a hard pitted stone and has a very strong flavour, similar to oil of wintergreen. The seeds were often ground up into cream and applied as medicine for the scalp. Quandong trees were thought to have other powers. The Madi Madi said that if you carved a hole in the quandong tree and placed something belonging to a person in it, that person would slowly waste away. The wood of the tree was sometimes used for firemaking using the saw method.[10]

They also illuminate another type of quandong, known as bitter quandong (*Santalum murrayanum*), which has brown to yellowish-red fruits

… very bitter to the taste. Some people were able to eat it nevertheless, and were much admired for this talent. Roasting was reported to remove the bitter taste. A stupefying drink called COOTHA was made from the root and bark at Lake Boga.[11]

The writer Jennifer Isaacs, who has researched and written extensively on Indigenous plant use across Australia, also recorded several other uses for quandong that she observed in Aboriginal communities in the Northern Territory:

Bark shavings are soaked, and liquid rubbed on itchy areas. Decoction of leaves and bark drunk as a purgative. Decoction of outerwood drunk for 'sickness of the chest'. Infusion of roots used for rheumatism and applied to the body for refreshment when hot and tired. Leaves are burnt to drive away mosquitos, and people 'smoke' themselves and their babies to gain strength for long trips.[12]

Colin Pardoe has worked extensively with Aboriginal groups across the Murray-Darling Basin. Rather than excavate archaeological sites, he looks at their spatial dimensions – how they are distributed across the landscape, how they are connected, and what that can tell us about how people lived, moved and used resources. For example, he and archaeologist Dan Hutton have shown how Barapa villages and hamlets cluster around ecological hotspots on the Murray River wetlands.[13] In studying quandong use north of the Murray, in Barkandji and Wiradjuri Country, he noted

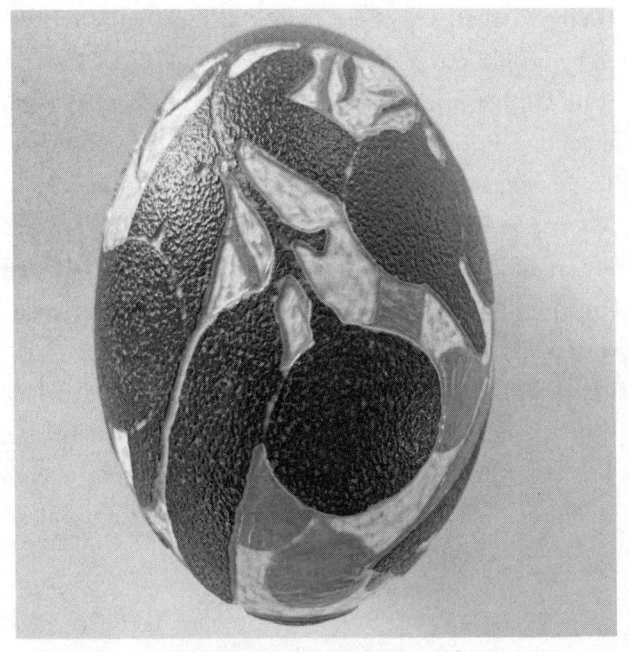

FIGURE 8.1: Carved emu egg featuring quandong by David Doyle, 2020.

that quandongs and emus share connections across the landscape. The spiritual and the ecological are closely entwined:

> Dreaming stories often reference a symbiotic relationship between quandongs and emus. The birds eat the fruit in large quantities, swallowing the nuts. Their stone-filled crops begin the digestive process so that as they wander the landscape they excrete a ready supply of quandong nuts whose germination is assisted by fertiliser ... Aboriginal people recognised the role of the emu in aiding their distribution.[14]

In the Kimberley a different species, *Santalum lanceolatum*, is not eaten by people and is dismissed as 'emu tucker' by Biddy Simon, our yam teacher from Chapter 6. The close link between all these different dimensions of landscape connection is encapsulated in the depiction of quandongs in the carved emu egg (Figure 8.1).

Carved by Barkandji artist David Doyle in 2020, the quandong in this artwork embodies the symbiotic relationship between emus and plants on Barkandji Country. Emus consume seeds from a vast array of plant species, including the quandong, and disperse them over long distances.[15] Further, the multi-layered holistic relationship between plants, people and animals is shown in this work, a celebration of deeply embedded aspects of subsistence knowledge within the cultural and spiritual world of Indigenous peoples.[16]

NUTCRACKERS

Quandong kernels are extremely valuable for food, medicine and cosmetics, but they are also a very hard nut to crack. Colin Pardoe worked with archaeologists Richard Fullagar and Ebbe Hayes to examine 1327 grinding implements from across the Murray-Darling Basin,[17] identifying twenty-four specialised implements used to crack and process quandong kernels.[18] (The grindstone story overall is a very important one the archaeologists are still working on – 'Given the paucity of stone sources throughout the Basin ... virtually all implements represent trade activity into local Aboriginal territories from elsewhere.')[19] While most Australian grindstones were used for multiple purposes, as discussed in the spinifex chapter, the quandong stones, or 'nutcrackers', as Wiradjuri elder Robert Clegg called them,[20] seem to have been very specialised.

As we also saw in the previous chapter, the patterns of wear and residues on the grindstones can help us understand what they were used for and how they were used. In this case they were used over long periods of time, and provide a material connection to generation upon generation:

> One of the attractions of these objects is the visible wear that is the result of ancestral elbow grease. The deep hollows in grinding dishes, the polish, cracks or pits in mortars, all provide tangible links down through the generations. Many of the wear facets can only be the result of continued use measured in centuries, not years or days.[21]

On one side the quandong grindstones have pits where the nuts were cracked, and on the other is a grinding surface where the flesh of the kernel was processed into an oily paste, once enough nuts had been cracked. Microscopic analysis of the residue 'showed abundant oils but no starch grains. This supports previous research that detected only trace amounts of starch in quandong kernels.'[22]

As with the emu connection, the quandong grindstones also tell a story about the broader use of Country. The trees are not spread evenly throughout the landscape but cluster in groves. Pardoe and colleagues tell the story:

Quandong stones are heavy enough that they become impractical to carry. The fact that they have several pits as well as a mortar bowl on the opposed surface suggests that they may have been kept at particular groves of quandongs, to be used when large numbers of seeds needed to be processed. This large-scale collecting strategy can be contrasted with an occasional expedient foraging strategy, where fortuitous collections of quandongs might be brought back to the settlement to be processed on a household mortar. During periods of greater mobility, people might encounter single quandong trees and could process fewer nuts with the portable mortars or kulki: foraging rather than collecting.[23]

Quandong trees are not randomly distributed within Aboriginal territories; they are clustered in groves in specific areas, fruit once a year at a regular time, and are harvested between June and October ... As one of the few sizable fruits available, Aboriginal people would visit these places on a seasonal basis to harvest the crop over several weeks. While the fruit is readily picked, dried fruit is of greater value as it can be ground or mashed and made into cakes that can be stored and transported long distances. Neither of these activities requires nut cracking but the mortar dish on the obverse face of the anvil would serve this purpose. Once the fruit had been removed, the focus would switch to cracking the nuts to extract the kernels which would be ground into ointment or salve for use through the year or to trade. While people might live close to the productive groves for some weeks at a time during the season, the quandong stones, being heavy, would be cached *in situ* until the following season.

Quandong stones indicate a foraging strategy of moving to the resources when in season for large scale processing.[24]

SUSTAINABILITY

As part of the Santalaceae family, quandongs are hemiparasitic, which means they need to source inorganic nutrients and water from the roots or branches of host plants.[25] These requirements, together with extensive land clearing and introduced species following European settlement, have made them endangered in many areas.[26]

A recent example of attempts to conserve and protect quandongs for the future comes from northern Victoria,[27] where Torrumbarry resident Tuesday Browell bought a piece of land containing a number of endangered trees, and put a conservation covenant on the property. She is now working with Wollithiga man Henry Atkinson to protect the plants and propagate new ones.

Why is it so important to try to conserve this plant? In this time of mass extinctions we need all the biodiversity we can get, especially plants adapted to increasingly arid conditions. But quandongs do something extra: they preserve biocultural history. They connect to Indigenous practices spanning millennia, and their wellbeing reflects the human interactions that have fostered their growth. That biocultural history also provides ideas and insights for a future in which traditionally Indigenous foods continue, and become more widely used across the Australian community.

9

RESPECTING KNOWLEDGE

As evidenced by the incredible longevity of Australia's First Peoples' many unique cultures, our scientific systems of land and cultural management were and are more than capable of sustaining people in the long term. In recent years there has been a huge surge in interest in what are commonly known as 'bush foods', the traditional foods of Australia's diverse Indigenous peoples. The explosion in interest and growth is a positive step, but the benefits aren't equitably shared. The complex networks of synergistic relationships between human, plant, animal and insect that make up Indigenous plant knowledge are often hugely simplified in their representation as 'bush tucker'. Further, while the 'bush food' industry does provide some important opportunities, shockingly, at present only around 1 per cent of the produce and the profits are generated by Aboriginal and Torres Strait Islander peoples and communities.[1] The recognition and protection of our knowledge, especially our knowledge of plants and our traditional foods, is a pressing problem on a number of fronts.[2]

Somehow our foods have come to be seen simply as seasoning or garnish, something just to add a little flavour. It's a small leap to imagine that the negative stereotypes covered in chapters 2 and 3 might well have contributed to this. My main gripe with this industry, apart from my outrage that our communities receive so little monetary and intellectual-property benefit, is that the deep knowledge of plants honed by our people is lost when our foods are seen as some type of novelty – a subtext I too often detect in the marketing and celebration of our foods in the 'bush tucker' realm. Where in this 'bush tucker' story do we foreground the recognition of our longstanding cultural practices, central to the development of scientific understandings and utilisation of our plants? Where do we tell the truth about the breadth and depth of the skills and innovation that have allowed us to work with Country and our plants in a sustainable way, ensuring everything we needed to be healthy and fulfilled over many millennia? Framing traditional foods as 'an accompaniment' perpetuates the idea that our fruits, vegetables, seeds and grains cannot form a whole diet, and within this is perhaps the continuation of ideas that we were a people and Country of little value, barely subsisting, just scraping together an existence before the Europeans came, the supposed architects of abundance.

This distortion – the sense that our foods are scarce, not robust or resilient enough to be seriously considered as a better way of feeding our nation and looking after Country while we are at it – runs throughout many of the conversations around the topic today. These ideas are a lie, and slowly non-Indigenous scientific testing is showing what we have always known: the plants of our Country are highly nutritionally valuable and are the perfect inclusion in the human diet to ensure health and vitality.[3] Our foods are not nutritionally inadequate; the knowledge used to produce them and to extract their value is not lesser. It is to the great detriment of the health of Indigenous people and of Country, indeed of all of Australian society, that our traditional foods are no longer widely available as a substantial part of our diets and shared culture today. The opportunity our plants offer and the breadth of knowledge behind their applications should not be obscured by narratives that do not allow us to see complexity and abundance, that fail to see the whole and the exciting potential they hold.

THE COMMERCIALISATION OF PLANTS – THE NATIVE AND BUSH FOOD INDUSTRY

A recent analysis of plant species identified as 'key' by the leading industry body Australian Native Foods and Botanicals estimated the market value of just thirteen 'priority species' at $81.5 million per year, with forecast growth to $160 million by 2025.[4] Considering this only encapsulates a mere fraction of the many thousands of plants utilised and refined by Australia's First Peoples for a multitude of uses, it is easy to see the huge potential of this fast-growing industry. However, the disparity between what we need (economic empowerment) and what we get (celebration of our traditional culture, respect for our people's knowledge and recognition of our land management practices) goes to the core of many of the problems faced by Australia's First Peoples, highlighting the pressing need for greater self-determination. How can it be that we only receive around 1 per cent of the benefit of this burgeoning market when it is entirely based on our knowledges and careful custodianship of plants over millennia?

There are many varied challenges and circumstances that have resulted in this lack of benefit, including the ongoing effects of colonisation, which have greatly limited land access and ownership by Indigenous people and communities, and a lack of opportunity and support, especially through access to capital.[5] There is inadequate government funding to enable the research and development needed for successful market engagement, and we require more robust recognition and protection of Indigenous cultural and intellectual property through law.[6]

While there are exciting opportunities offered by the development of the botanicals and bush foods sector, the true benefit will not be realised without significant assistance to meet social, cultural and legal challenges. Greater support of Indigenous leadership will help build the necessary foundation of self-determination in the growth of the bush foods and native botanicals industry. Many Indigenous-led groups and governing bodies are emerging, ensuring greater potential to increase sustainable cultural and economic opportunities for Indigenous communities. Indigenous-led bodies have the capacity to expand the direct benefit of this industry for Indigenous communities and peoples through the empowerment of Indigenous-led research and development opportunities and culturally appropriate business models.[7] Several new governing bodies have recently been established, including the Northern Australia Aboriginal Kakadu Plum Alliance (NAAKPA), a consortium of northern Australian Aboriginal businesses focused on supporting each other to ethically harvest and process Kakadu plum on their homelands.[8] Initiatives to link culture and business to achieve economic benefit are supported by entities such as the Federation of Victorian Traditional Owner Corporations, which have recently established the Victorian Traditional Owner Native Food and Botanicals Strategy. This strategy was developed through workshops with community members to ensure culturally grounded design. The overarching aspiration is to empower traditional owners to successfully participate in the native food and botanicals industry through embedding culturally appropriate principles, such as restoring and reclaiming knowledge and working to improve, in tandem, the health and sustainability of Country, industry practice and operations.[9]

Another excellent example of Indigenous-led industry bodies is the First Nations Bushfood & Botanical Alliance Australia, which is managed by a board of Indigenous bush food professionals working towards substantially increasing Indigenous participation in the industry. One of the key actions the alliance is focusing on is implementing protocols that would set national standards on how to work with First Nations people in the industry, such as:

- Provenance and authenticity: to protect First Nations producers, respect our protocols and recognise our custodianship.
- Changes to the law: to respect and protect First Nations knowledge in bush foods and bush products. This includes intellectual property rights and penalties for misappropriation, and implementation of the Nagoya Protocol on Access and Benefit Sharing.
- Education and awareness: promoting respect for our First Nations knowledge, values and protocols.[10]

These foundational aspirations highlight an important aspect of the barriers that will need to be overcome in order for Aboriginal and Torres Strait Islander people to equitably participate in the market – the pressing need for knowledge protections through law.[11]

INDIGENOUS CULTURAL AND INTELLECTUAL PROPERTY RIGHTS AND BIODISCOVERY

It is always interesting to me that when I talk to people about Indigenous cultural and intellectual property (ICIP) rights relating to plant knowledge – a vital means through which to achieve greater self-determination and empower our communities – very often someone will take issue and say something along the lines of 'Aboriginal people can't *own* plants'. This is striking when we consider that very few voice any opposition to huge, faceless multi-billion-dollar corporations being able to 'own' plants with the patents they take out on them. These companies very often use, without any attribution, knowledge taken from Indigenous peoples and communities while marketing the apparent 'discovery' of the plant's uses and properties as their own.[12] It is a common misconception that Indigenous peoples should not be entitled to any monetary reward or ICIP rights for the resources they have developed over time, as if we ourselves are supposed to be a part of nature, requiring no income or rights to live or to keep our communities and knowledge strong. In order for our communities to remain sustainable, we need to participate in mainstream economies, and the idea that our rights are not equal to those of corporations and businesses is absurd.

Worse still, recently some non-Indigenous companies have been allowed to patent traditional medicinal plants using their Indigenous names – so they not only own the plant but the word too. This means that the Indigenous community from which the plant and its language name have come now cannot use their own language or the plant, which has been a part of their culture for countless generations, for commercial gain. This was the case for the *gumbi gumbi* plant (*Pittosporum angustifolium*), successfully patented for use by a non-Indigenous company, effectively denying any financial benefit to Indigenous communities while also denying knowledge rights, preventing them from using, trading and exporting their own culturally and historically important plant.[13] Australian intellectual property and access- and benefit-sharing laws currently allow outsider companies and individuals to achieve private ownership of Indigenous resources, using their knowledge for commercial purposes and excluding others, including the Indigenous communities from which their use was gleaned, from the market. These outside entities are taking ownership of these plants – and the knowledge that is embedded in their use – through legal licences such as patents. This patenting landscape sits within what is termed 'biodiscovery'. The commercial exploitation of biochemical or genetic material, especially by placing controls on its usage and application while failing to pay fair compensation to the communities from which it originates, is known as 'biopiracy'. Biopiracy is a significant concern in Australia, as informed consent and the attribution of ICIP rights are often entirely missing, with benefit sharing and acknowledgement of pre-existing Indigenous knowledge also absent in the many hundreds of patent applications lodged.[14]

While there has been some recognition of the ICIP rights of Indigenous communities and the need for protections in the realm of biodiscovery and biopiracy, protections are still limited and there has as yet been no national approach to this problem.[15] As long as there are no uniform laws across the nation to acknowledge and protect Indigenous knowledge, there will remain significant barriers to our equitable participation in the bush foods and botanicals market. Our knowledge will continue to be appropriated and co-opted, denying our communities opportunities to participate in the market and denying future generations their rightful place within the economy, as well as their custodial right to use and to benefit from the knowledge of their people, their inheritance.

FUTURE FOODS AND CLIMATE CHANGE

While showing huge potential in market value, Indigenous plants still remain relatively unrecognised. In its 2019/20 market study, the Australian Native Foods and Botanicals peak body reported that

> Australia accounts for around 10 per cent of the world's biodiversity and has extensive arable landscapes, yet the macadamia (*Macadamia integrifolia*) is the only native horticultural species cultivated at scale. European settlers have tended to focus on cultivating familiar Northern Hemisphere species.[16]

In Australia today, we are largely ignoring foods that are perfectly suited to our environment and growing many that Country must pay a high price to nurture. In 2020, our biggest horticultural export was almonds, an imported food that is highly detrimental to Country and to biodiversity.[17] In 2019/20 this crop generated $772.6 million.[18] Grown as a monoculture, almond crops greatly deplete biodiversity and require extraordinary amounts of water to grow – between 8.5 and 10 megalitres per hectare, sometimes rising to 14 depending on climatic variables, roughly three times more water than is needed to grow crops such as wheat.[19] Almond crops rely heavily on the use of bees, while also having been shown to cause devastating mass bee deaths through their notable dependency on harmful pesticides.[20] Alarmingly, the share of land devoted to this crop has grown by 50 per cent since 2016 within the Murray-Darling Basin alone, a geographical area that has been stricken with water scarcity issues in recent years.[21] It is clear that the foods we grow will need much more careful consideration, especially as we face pressures such as drought and other catastrophic events caused by climate change. For too long we have allowed crops that are a drain on Country to expand, often across areas already strained. The Murray-Darling Basin is one such example, pushed to its limits through the expansion of almonds, cotton and other industries that, while lucrative for some, take far too much from Country.

The agricultural industry and food security will increasingly be threatened by changes in the environment. Many of the introduced grains that the Australian food industry is heavily reliant on, such as wheat, are likely to be significantly compromised as a result of the pressures of climate change. There are many plants indigenous to Australia that can cope with the climatic fluctuations predicted. For example, indigenous grasses are highly adaptable and resilient to the multiple pressures of climate change, especially drought and fire. Indigenous grasses, such as native millet (*Panicum decompositum*) and kangaroo grass (*Themeda triandra*), have been used extensively and effectively for food by Australia's diverse Indigenous peoples for many millennia. Further, unlike many introduced food crops, native grasses have substantial capacity to offer dual environmental and economic benefits, as they are able to sequester carbon and promote biodiversity.[22]

The march of invasive species is also a huge pressure on Country, and will greatly affect the plants we are able to grow and enjoy in the future. For example, many native grasses are catastrophically compromised by buffel grass (*Cenchrus ciliaris*) and white buffel grass (*C. pennisetiformis*), aggressive weeds that overrun native grass ecosystems, converting them from havens of biodiversity into monocultures. This harms native wildlife by limiting food and shelter options, and buffel grasses in particular also greatly impede the effectiveness of cultural fire regimes, making fires burn too hot. Addressing the devastating impact of invasive weeds and species on biodiversity must be a part of the conversation if we are to strengthen and expand our effective use of native plants.[23]

Further research and incorporation of native crops into our current agricultural systems may be an important factor in ensuring future food security in Australia. While native grasses remain unproven as viable large-scale crops, existing large-scale agriculture and monocultures have proven highly damaging to the environment, and it is therefore salient to consider indigenous plant foods as not only being viable as alternative foods but also as an alternative to industrial-scale farming. These crops also have the potential to provide expanded opportunities for Indigenous Australians, empowering their knowledge and land management practices in food production and farming.[24]

HEALTH AND WELLBEING – PEOPLE AND COUNTRY

Research has shown that there are substantial improvements in health outcomes and wellbeing for Indigenous Australians who have access to their traditional foods.[25] But this is rarely available to most Indigenous people in Australia, especially those who live in urban areas. The loss of access to Country to cultivate and harvest traditional foods was one of the very first destructive effects of colonisation, posing great threats to our health and continuing to do so today.[26] Across Australia, many Indigenous organisations are undertaking projects to produce our foods and make them available once again, reinvigorating traditional knowledge and land management. An excellent example is Uncle Bruce Pascoe's Black Duck Foods, which is discussed in more depth in *Country*, his co-authored addition to the First Knowledges series. Projects that resource, empower and expand the cultivation of traditional indigenous grasses and other native food crops have the potential to be beneficial in both urban and remote areas, and have the further advantage of providing significant opportunities for Indigenous peoples to work on Country, which, like having access to our foods, has also been shown via a wealth of research to be highly beneficial to Indigenous health and welfare.[27] These ventures also provide further opportunities to reinvigorate traditional cultural fire practice, a highly effective land management method with a multitude of benefits, including the revitalisation of grasslands and biodiversity across Australia.[28]

Native plants traditionally used by Indigenous peoples are perfectly suited to the often harsh conditions of the Australian environment. They do not require fertilisers or pesticides, use little water, are highly nutritionally valuable and, as discussed, many have been shown to cope with climate fluctuations much better than introduced species.[29] Opportunities to study, produce and reincorporate native grasses in particular into the Australian diet show much promise in substantially benefiting both Indigenous and non-Indigenous Australians, as well as biodiversity and the environment in general.[30] The re-establishment of our native foods may also provide opportunities for Indigenous Australians and our knowledge and land management practices to become a significant part of food production and farming in Australia.

However, this cannot happen without the empowerment of principles of self-determination, especially through greater protection of our plant knowledge within the law. So much has been taken from us as a result of the ongoing circumstances of colonisation; benefit sharing through protections in law will enact reparations to people and to Country. All Australians lose when big, predominantly overseas companies are able to take our plants and the knowledge around their use from us. Giving our knowledge robust legal protections will strengthen them for today and for communities yet to come, and allow us much-needed opportunities for self-determination that have been missing, across the board, for too long.

10

FUTURES

Plants signify abundance. Anyone who has ever planted, propagated, watered, harvested or simply watched them grow has recognised the wealth and potential they embody. Humans rely on them for so much, yet like so many things that are fundamental to our survival, we too often take them for granted.

Australian indigenous plants offer a magnificent portal through which all Australians can learn of the rich knowledge and complex land management systems of Aboriginal and Torres Strait Islander peoples. Their effectiveness as a gateway to greater understanding and respect is very evident, for example, in the interest young children often show in planting, growing and connecting with indigenous plants, learning of their uses over time and the profound knowledge embedded within them. Expanding resources and opportunities to educate about plants, especially for our future generations, ensures that we are providing the building blocks for connection to Country for all. It is only through nurturing deep connections that we can fully accommodate our responsibilities as custodians.[1]

This connection is what is missing from modern Australia. You can see it in the obsession with European plants and gardens. In the types of plants we prop up at the expense of unsustainable water consumption and fertiliser use. Practices that hurt Country. An obsession that blinds modern Australia to the riches of this continent and to the potential for agricultural production that promotes healthy Country and healthy people. The agency and innovation of Aboriginal people and their use of plants demonstrated in this book are but a glimpse into the plants that have sustained more than 65,000 years of civilisation and biodiversity.

For too long Australian society has missed opportunities to deeply connect with Country and activate this necessary custodial obligation to ensure her current and future health. Many of our failures in effecting this meaningful connection come from an inability to engage wholeheartedly in necessary truth-telling. We seem as a society to be hovering above Country, never deeply engaged enough within her, almost as though we are too scared to really look her in the eye, and to find where we each fit. It feels like she is being driven like she's been stolen, thrashed and pushed to her limits.

The first step towards a more sustainable future is to open our eyes. How can we move forward if we don't know which way we are facing? The blindness of modern Australia is one born from a wilful ignorance that stems from our racist past and present. Despite vigorous longstanding and ongoing challenges to this wilful blindness by activists, academics and elders, modern Australia largely continues to lie to itself about its history and the history of this land. This continent has been cared for and loved continuously for longer than any other place on Earth. People and land depend on each other. Indigenous people have shaped this land and have kept it and themselves healthy for tens of thousands of years, despite massive shifts in global climate and sea level. We hold the manual for a safe and sustainable life on this unique landmass. But to read it, we must first be able to see. We must open our eyes. We must see the truth.

Truth is so fundamental that it is the almost universal lesson we parents try to impart to our children. To tell the truth. No matter how bad a situation gets, telling the truth is the first step required to truly resolve past harms. The flip side of this coin is the biblical adage 'lies beget lies'. The irony of the source of this truism is not lost on us. Nevertheless, it rings true. True healing and connection for the Australian people of today and our Country requires unfettered and relentless truth-telling at every level of our society. Only then can we make reparations and begin to heal. Only then can we form trusting reciprocal partnerships that allow us to truly come together and utilise the many tools at our disposal in order to meet the challenges of our future.

For more than 200 years we have seen how ultimately ineffective extractive modes of engagement are leading to short-term gains, long-term distrust and irreversible damage. There is one thing we have never tried – the widespread empowerment of Indigenous communities to care for Country, to speak for Country, to be in the driver's seat.[2]

In Australia, one of the first things to be denied to Indigenous communities through the violence of colonisation was access to Country, to our plants and traditional foods.[3] This had catastrophic impacts and many were killed simply trying to maintain traditional practices to feed their communities and to fulfil their spiritual and cultural obligations. Respectful management systems that were developed and refined through careful observation, scientific testing, reciprocity and applied knowledge, a vast knowledge system of the living and non-living components of our world, became inaccessible. We once had what we needed to be healthy. We understood the rhythms of our Country. Now, many of us live with the ongoing legacy of being forced off Country and forced to be dependent on the invaders for sustenance.

The aspiration of Indigenous food sovereignty is becoming a recognisable movement in Australia, with many communities beginning to test and engage in the production of traditional foods and the reinvigoration of traditional land management practices.[4] The federal government recently expanded support for 'working on Country' ranger programs,[5] and representative Indigenous bodies are increasingly being given funding opportunities to pursue programs of land management, particularly related to biodiversity and engagement with native foods.[6] However, we are still very much behind the Indigenous food sovereignty movements in other countries (such as the United States), and much will have to change before we see the day that the majority of our people again have access to their Country and traditional foods. Despite the known positive outcomes for health connected to access to the foods our ancestors have eaten for millennia, our people are still too often locked out of ancestral homelands, and having access to our foods is not explicitly recognised, for example, in the Closing the Gap initiative. In real terms, it is difficult to see how Indigenous food sovereignty may be effected without greatly expanded ownership of lands, and the many Indigenous peoples who live in urban areas must not be left out.[7]

The challenges we face may, at times, seem overwhelming, but we can produce so much benefit in the small things we are able to do. We also have the power as people to come together to effect massive change. Being pessimistic and overwhelmed will not light the path forward, but insistent truth-telling in every part of our lives through self-education, through reading and listening to the words of Indigenous peoples and, most of all, empowering their ways of doing, and a whole lot of individual and collective action to nurture Country, will go a long way.

Together we must consider that which largely remains hidden when we weigh the wealth of this place, Australia, in the context of our current, multiple crises. Could it be that when the colonisers came to pillage the riches, the gold, the soil, the lands, they were ignorant to the most important wealth of all: the Indigenous worldview. There is time yet to avoid the irrevocable loss of so much we hold dear, to enjoy the spoils of the true wealth that comes from nurturing Country, by following the Law set down and followed by our Old People over thousands of generations. When you look after Country, your mother, she will look after you.

It is not too late.

ACKNOWLEDGEMENTS

Huge thanks to Aunty Sarah Martin, Uncle Badger Bates and David Doyle, who gave much of their time and expertise to help me with many aspects of this book.

Thanks also to Sally Heath for expert guidance and to Sam Palfreyman for polishing my work so skilfully. Thank you Aunty Joy Murphy Wandin, Aunty Vicki Couzens, Lesley Head, Michael-Shawn Fletcher, Krystal De Napoli, Stacie Piper, Genevieve Grieves, Shannon Faulkhead, Adrianne Semmens, Raymond Zada, Rachel Carey, Jo Henry, Matt Rice, David Summers, Darren Griffin, Uncle Bruce Pascoe, Jarrod Hughes, Eddie Cubillo, Lyndon Ormond Parker, Terri Janke, Helen Taylor, Fiona Cornforth, Loren Hackett and Kate Howell, for encouraging and supporting me and being the best, super-brainiest friends available to humanity. Special thanks to all the Aboriginal and Torres Strait Islander community members that I have been honoured to work alongside, but especially those who contributed directly to this book, yarning with me and generously giving permission for me to share their knowledge and words: Uncle Badger and David, as well as Jonathan Jones, Jessyca Hutchens, Uncle Brendan Kennedy, my tidda Maddison Miller, Mark J Grist and Uncle Rodney Carter.

Lastly, to all my mad family, immediate and extended: I love you. Extra special thanks to my sister Nici Cumpston, who guides and backs me in all I do, and the biggest love and appreciation to my beautiful fellas, my everything – Damien, Lou and Hector. – ZC

My debt to a wide range of researchers will be apparent from the citations in the text. Particular thanks must go to my plant teachers in the East Kimberley, Mrs Biddy Simon and the late Mrs Polly Wandanga. Jenny Atchison and Richard Fullagar have also been indispensable long-term companions on the planty journey. Finally, thanks also to Sally Heath for her patience and encouragement.
– LH

I would like to acknowledge the Wurundjeri, who gave me permission to work on their Country and who generously shared their knowledge with me. Thanks also to all my Aboriginal mentors and colleagues who have shaped my contributions to this book. Lastly, thanks to Nicki, who is my rock and my sounding board and who I would be lost without. – MF

IMAGE CREDITS

Inside covers	*murrum (overflowing)* from the 'Emu Sky' Exhibition, University of Melbourne © Aunty Kim Wandin, Simon Briggs, David Doyle, Jonathan Jones, Aunty Julie Freeman, Lachlan McDaniel, Uncle Roy Barker Snr, Nici Cumpston, and an unknown south-eastern Aboriginal artist 2021 Photograph by Christian Capurro
p.31	Uncle Badger Bates on Country, teaching about digging sticks © Zena Cumpston 2021 Photograph by Zena Cumpston
p.52	Map of Bolin Bolin Billabong © Casey Schuurman 2022
p.70	Mary, Jacob and Doughboy on Barkandji Country Photograph by Frederic Bonney c. 1879. Mitchell Library, State Library of NSW, PXA 562.
p.101	Schematic diagram of a *Typha* rhizome and the string-making process © Casey Schuurman 2022 Reproduced by Casey Schuurman from an original illustration by Hannah Fullagar
p.155	An emu egg with quandong carving © David Doyle 2022 Photograph by Zena Cumpston

NOTES

INTRODUCTION
1. S Anna Florin et al., 'The First Australian Plant Foods at Madjedbebe, 65,000–53,000 Years Ago', *Nature Communications*, 11, 17 February 2020, <doi.org/10.1038/s41467-020-14723-0>.

1. PERSONAL PERSPECTIVES
1. Uncle Badger Bates, 'Statement for the Murray-Darling Basin Royal Commission', 2018, <environment.sa.gov.au/files/sharedassets/public/river_murray/royal-commission/submissions/william-badger-bates-barkandji-nsw-mdb-rc-gen.pdf>.
2. Zena Cumpston, *Indigenous Plant Use: A Booklet on the Medicinal, Nutritional and Technological Use of Indigenous Plants*, Clean Air and Urban Landscapes Hub, University of Melbourne, 2020, <nespurban.edu.au/wp-content/uploads/2020/08/Indigenous-plant-use.pdf>.
3. For more information, go to the Australian Government's State of the Environment (SoE) reporting page, <awe.gov.au/science-research/soe>.
4. For a readable history of Australian archaeology, see Billy Griffiths, *Deep Time Dreaming: Uncovering Ancient Australia*, Black Inc., Melbourne, 2018.
5. It was not only Australian evidence that brought progressivist ideas of human evolution undone. Northern hemisphere conceptual frameworks have also changed significantly since that time.
6. Lesley Head, *Second Nature: The History and Implications of Australia as Aboriginal Landscape*, Syracuse University Press, New York, 2000, pp. 150–1.
7. I have written about this in more detail in my book *Second Nature* (see previous note); see in particular pp. 133–7.
8. Beth Gott, 'Cumbungi, *Typha* Species: A Staple Aboriginal Food in Southern Australia', *Australian Aboriginal Studies*, 1, 1999, pp. 33–50 (p. 42).
9. Lesley Head, 'The Holocene Prehistory of a Coastal Wetland System: Discovery Bay, South-Eastern Australia', *Human Ecology*, 15, 1988, pp. 435–63; Lesley Head, 'Holocene Vegetation, Fire and Environmental History of the Discovery Bay Region, South-Western Victoria', *Australian Journal of Ecology*, 13, 1988, pp. 21–49.
10. Polly has since passed away, but I use her name with the permission of her family.

11 Example publications from this project include Lesley Head & Richard Fullagar, '"We All La One Land": Pastoral Excisions and Aboriginal Resource Use', *Australian Aboriginal Studies*, 1, 1991, pp. 39–52; Lesley Head, 'Aborigines and Pastoralism in Northwestern Australia: Historical and Contemporary Perspectives On Multiple Use of the Rangelands', *The Rangeland Journal*, 16, 1994, pp. 167–83; Richard Fullagar, David M Price & Lesley Head, 'Early Human Occupation of Northern Australia: Archaeology and Thermoluminescence Dating of Jinmium Rockshelter, Northern Territory', *Antiquity*, 70, 1996, pp. 751–73; Lesley Head & Richard Fullagar, 'Hunter-Gatherer Archaeology and Pastoral Contact: Perspectives from the Northwest Northern Territory, Australia', *World Archaeology*, 28(3), 1997, pp. 418–28; Lesley Head, Jennifer Atchison & Richard Fullagar, 'Country and Garden: Ethnobotany, Archaeobotany and Aboriginal Landscapes near the Keep River, Northwestern Australia', *Journal of Social Archaeology*, 2, 2002, pp. 173–96.
12 Lesley Head, 'Landscapes Socialised by Fire: Post-Contact Changes in Aboriginal Fire Use in Northern Australia, and Implications for Prehistory', *Archaeology in Oceania*, 29, 1994, pp. 172–81; Lesley Head, 'Rethinking the Prehistory of Hunter-Gatherers, Fire and Vegetation Change in Northern Australia', *The Holocene*, 6, 1996, pp. 501–7.
13 Jennifer Atchison & Lesley Head, 'Yam Landscapes: The Biogeography and Social Life of Australian *Dioscorea*', *The Artefact*, 35, 2012, pp. 59–74.
14 Jennifer Atchison, 'Human Impacts on *Persoonia falcata*: Perspectives on Post-Contact Vegetation Change in the Keep River Region, Australia, from Contemporary Vegetation Surveys', *Vegetation History and Archaeobotany*, 18(2), 2009, pp. 147–57.
15 Ethnobotany is the study of cultural plant use and perception of plants. It is often associated with traditional and local knowledge.
16 Lesley Head & Pat Muir, *Backyard: Nature and Culture in Suburban Australia*, University of Wollongong Press, Wollongong, 2007; Lesley Head & Pat Muir, 'Suburban Life and the Boundaries of Nature: Resilience and Rupture in Australian Backyard Gardens', *Transactions of the Institute of British Geographers*, New Series 31(4), 2006, pp. 505–24.

2. LOOKING BACK, MOVING FORWARD

1 Uncle Brendan Kennedy, in a presentation given as part of a University of Melbourne Law School NAIDOC event, 'Water Sovereignty, Climate Futures and the Academy: Can We Heal Country on Stolen Indigenous

Land?', 14 July 2021, <law.unimelb.edu.au/about/welcome/mls-indigenous/home/iljh/resources/videos/water-sovereignty>.
2 Uncle Brendan Kennedy.
3 Kirsten Parris, Briena Barrett, Helaine Stanley & Joe Hurley (eds), *Cities for People and Nature*, Clean Air and Urban Landscapes Hub, Melbourne, 2020, <nespurban.edu.au/wp-content/uploads/2020/11/Cities-for-People-and-Nature.pdf>.
4 George Nicholas, 'It's Taken Thousands of Years, but Western Science is Finally Catching Up to Traditional Knowledge', *The Conversation*, 15 February 2018, <theconversation.com/its-taken-thousands-of-years-but-western-science-is-finally-catching-up-to-traditional-knowledge-90291>.
5 Nelly Zola, Beth Gott & Koorie Heritage Trust, *Koorie Plants, Koorie People: Traditional Aboriginal Food, Fibre and Healing Plants of Victoria*, Koorie Heritage Trust, Melbourne, 1992, pp. 27–8.
6 Penny Olsen & Lynette Russell, *Australia's First Naturalists: Indigenous Peoples' Contribution to Early Zoology*, National Library of Australia, Canberra, 2019, p. 9.
7 Harry Allen (ed.), *Australia: William Blandowski's Illustrated Encyclopaedia of Aboriginal Australia*, Aboriginal Studies Press, Canberra, 2010.
8 Philip A Clarke, *Aboriginal Plant Collectors: Botanists and Australian Aboriginal People in the Nineteenth Century*, Rosenberg Publishing, Dural, 2008, p. 9.
9 Mark Dugay-Grist, 'Time Separates People, Knowledge Binds People', in Allen (ed.), *Australia: William Blandowski's Illustrated Encyclopaedia of Aboriginal Australia*, p. 2.
10 Erin Matchan & David Phillips, 'Victoria's Volcanic History Confirms the State's Aboriginal Inhabitation Before 34,000 Years', *Pursuit*, n.d., <pursuit.unimelb.edu.au/articles/victoria-s-volcanic-history-confirms-the-state-s-aboriginal-inhabitation-before-34-000-years>; see also Erin Matchan, David Phillips, Fred Jourdan & Korien Oostingh, 'Early Human Occupation of Southeastern Australia: New Insights from 40ar/39ar Dating of Young Volcanoes', *Geology*, 48(4), 2020, pp. 390–4, <doi.org/10.1130/G47166.1>.
11 Hugo Schulz & Ferdinand von Mueller (trans., rev.), *Excerpts from Professor Hugo Schulz's Treatise on Eucalyptus Oil; Reprinted from* The Australasian Medical Gazette *for 1883*, L Bruck Medical Publisher, 1883, p. 4, <archive.org/details/ExcerptsfromPro00Muel/mode/2up>.
12 Schulz & von Mueller, p. 7.

NOTES

13 Maddison Miller, 'Blak Emu', *Australian Archaeology*, 87(3), 2021, pp. 318–19, <doi.org/10.1080/03122417.2021.1991442>.

14 Michael-Shawn Fletcher, Anthony Romano, Simon Connor, Michela Mariani & Shira Yoshi Maezumi, 'Catastrophic Bushfires, Indigenous Fire Knowledge and Reframing Science in Southeast Australia', *Fire 2021*, 4(3), 61, <doi.org/10.3390/fire4030061>.

15 Ian J McNiven, 'Bandwagons and Bathwater', *Australian Archaeology*, 87(3), 2021, pp. 316–17, <doi.org/10.1080/03122417.2021.1991436>.

16 Museum of Victoria Aboriginal Studies Dept, *Women's Work: Aboriginal Women's Artefacts in the Museum of Victoria*, Museum of Victoria Aboriginal Studies Dept, Melbourne, 1992, p. 1.

17 Lynette Russell, *Savage Imaginings: Historical and Contemporary Constructions of Australian Aboriginalities*, Australian Scholarly Publishing, Melbourne, 2001, p. 39.

18 Martuwarra RiverofLife et al., 'Yoongoorrookoo: The Emergence of Ancestral Personhood', *Griffith Law Review*, 2021, p. 26, <doi.org/10.1080/10383441.2021.1996882>.

19 Chelsea Watego, 'Always Bet on Black (Power)', *Meanjin*, Spring 2021, <meanjin.com.au/narratives/8-chelsea-watego-always-bet-on-black-power/>.

20 Ross L Jones, 'Eugenics in Australia: The Secret of Melbourne's Elite', *The Conversation*, 21 September 2011, <theconversation.com/eugenics-in-australia-the-secret-of-melbournes-elite-3350>.

21 Luke Pearson, 'Indigenous Science – Setting the Record Straight', *IndigenousX*, 1 November 2018, <indigenousx.com.au/indigenous-science-setting-the-record-straight/>; see also Ross L Jones, *Humanity's Mirror: 150 Years of Anatomy in Melbourne*, Haddington Press, Melbourne, 2007, pp. 8, 117.

22 Cressida Fforde, 'Collecting, Repatriation and Identity', in Cressida Fforde, Jane Hubert & Paul Turnbull (eds), *The Dead and Their Possessions: Repatriation in Principle, Policy and Practice*, Routledge, Taylor & Francis Group, New York and London, 2002, pp. 25–46, (p. 32).

23 Fforde, p. 37.

24 Mick O'Loughlin, 'Kooriculum Indigenous Science Program', *IndigenousX*, 30 July 2020, <indigenousx.com.au/kooriculum-indigenous-science-program/>.

25 Djandak, *Kalimna Park Management Plan 2021–2026 Consultation Draft*, 12 April 2021, <djandak.com.au/wp-content/uploads/2021/05/Kalimna-Park-Management-Plan-2021-2026-Consultation-Draft-V03-20210412.pdf>.

26 Joseph Dunstan, 'After 160 Years, Aboriginal Cultural Burning Returns to Coranderrk Station', *ABC News*, 21 April 2021, <abc.net.au/news/2021-04-21/aboriginal-cultural-burning-at-coranderrk-station-wurundjeri/100082382>; see also 'Park Lands Cultural Burn', City of Adelaide website, <living.cityofadelaide.com.au/cultural-burn-park-lands/>.

27 'City Adopts Cultural Development Plan 2019–2029', City of Perth website, <perth.wa.gov.au/news-and-updates/all-news/city-adopts-cultural-development-plan-2019-2029>.

28 'Yarra River Protection (Wilip-gin Birrarung murron) Act 2017', Victoria State Government website, <water.vic.gov.au/waterways-and-catchments/protecting-the-yarra/yarra-river-protection-act>.

3. BOLIN BOLIN

1 In conversation with Uncle Dave Wandin.

2 Jack Banister, 'Beneath Modern Melbourne, A Window Opens into Its Ancient History', *The Guardian*, 26 December 2019, <theguardian.com/australia-news/2019/dec/26/beneath-modern-melbourne-a-window-opens-into-its-ancient-history>.

3 James Boyce, *1835: The Founding of Melbourne & the Conquest of Australia*, Black Inc., Melbourne, 2011.

4 Boyce.

5 Boyce.

6 Anita Smith et al., 'Indigenous Knowledge and Resource Management as World Heritage Values: Budj Bim Cultural Landscape, Australia', *Archaeologies*, 15, 2019, <doi.org/10.1007/s11759-019-09368-5>.

7 Boyce.

8 Deborah Bird Rose, *Reports from a Wild Country: Ethics for Decolonisation*, University of New South Wales Press, Sydney, 2004.

9 George Orwell, *1984*, Penguin, Melbourne, [1949] 2011, p. 284.

10 AP Elkin, 'Our Aboriginal Problem', *The Australian Intercollegian*, 1 December 1933, p. 154.

11 See Uncle Bruce Pascoe's rebuff to one notable instance of this kind of categorisation in 'Andrew Bolt's Disappointment', Griffith Review, 36, April 2012, <griffithreview.com/articles/andrew-bolts-disappointment>.

12 Boyce.

4. ABUNDANCE

1 Takashi Kuribayashi, <takashikuribayashi.com/works>.
2 You may have seen Barkandji spelt in different ways, including Paakantyi and Barkindji – it is important to note that there is no one 'right' way to spell Aboriginal names as our languages were not written. I choose to use the spelling 'Barkandji', as it foregrounds our belonging to the Barka (Darling River, sometimes spelt Baaka). I know many of my mob choose different spellings, including David Doyle, who features in this chapter and prefers the spelling 'Barkindji'.
3 Frederic Bonney, 'On Some Customs of the Aborigines of the River Darling, New South Wales', *The Journal of the Anthropological Institute of Great Britain and Ireland*, 13, 1884, pp. 122–37 (p. 125), <doi.org/10.2307/2841717>.
4 Jeannette Hope & Robert Lindsay, *The People of the Paroo River: Frederic Bonney's Photographs*, New South Wales Dept of Environment, Climate Change and Water, Sydney, 2010.
5 Jeremy Beckett, Luise Hercus, Sarah Martin & Claire Colyer (ed.), *Mutawintji: Aboriginal Cultural Association with Mutawintji National Park*, Office of the Registrar, *Aboriginal Land Rights Act 1983* (NSW), Glebe, 2008.
6 This kind of knowledge transfer is explored through the lens of *mura* stories in Jeremy Beckett & Luise Hercus, *The Two Rainbow Serpents Travelling: Mura Track Narratives from the 'Corner Country'*, ANU Press Canberra, 2009, p. 10.
7 Nelly Zola, Beth Gott & Koorie Heritage Trust, *Koorie Plants, Koorie People: Traditional Aboriginal Food, Fibre and Healing Plants of Victoria*, Koorie Heritage Trust, Melbourne, 1992, p. 15.
8 Richard Fullagar & Judith Field, 'Pleistocene Seed-Grinding Implements from the Australian Arid Zone', *Antiquity*, 71, 1997, pp. 300–7, <doi.org/10.1017/S0003598X00084921>.
9 S Anna Florin et al., 'The First Australian Plant Foods at Madjedbebe, 65,000–53,000 Years Ago', *Nature Communications*, 11, 17 February 2020, <doi.org/10.1038/s41467-020-14723-0>; a precis of the study can also be found at <stories.uq.edu.au/news/2020/ancient-plant-foods-discovered-in-arnhem-land/index.html>.
10 Harry Allen, 'The Bagundji of the Darling Basin: Cereal Gatherers in an Uncertain Environment', *World Archaeology*, 5(3), 1974, pp. 309–22 (p. 313), <doi.org/10.1080/00438243.1974.9979576>.

11 Personal communication with Uncle Badger Bates on the Barka at Menindee, April 2021.
12 Allen, p. 315.
13 Bonney, p. 125.
14 Allen, p. 310.
15 Uncle Stan Grant Senior AM is a Wiradjuri elder and knowledge holder. Uncle Stan has long been a powerful figure in the Wiradjuri community, particularly for his tireless work in language revival.
16 Jonathan Jones, soundscape designed for *Emu Sky* [exhibition], Old Quad, University of Melbourne, November 2021–July 2022, <emusky.culturalcommons.edu.au/works/uncle-stan-jonathan-jones-grindstones/>.
17 Coolamon is derived from the Kamilaroi language word *gulaman*. While there are more than 250 diverse Indigenous languages, with many dialects, there are many words that have, since colonisation began, come into widespread use in Aboriginal English. Not all peoples refer to these items as 'coolamon', with many groups still using their own traditional language names. I use coolamon, as this is common on my Country and across south-eastern Australia more generally.
18 Philip A Clarke, *Australian Plants as Aboriginal Tools*, Rosenberg Publishing, Kenthurst, 2012, p. 28.
19 Museum of Victoria Aboriginal Studies Dept, *Women's Work: Aboriginal Women's Artefacts in the Museum of Victoria*, Museum of Victoria Aboriginal Studies Dept, Melbourne, 1992, pp. 1, 36–7.
20 Local Land Services Western Region, *Ecological Cultural Knowledge – Paakantyi (Barkindji): Knowledge Shared by the Paakantyi (Barkindji) People*, Local Land Services Western Region, 2016, pp. 43, 60, 65, <lls.nsw.gov.au/__data/assets/pdf_file/0005/737627/Paakantyi_Booklet_WEB-updated.pdf>.
21 Harry Allen (ed.), *Australia: William Blandowski's Illustrated Encyclopaedia of Aboriginal Australia*, Aboriginal Studies Press, Canberra, 2010, pp. 74–5.
22 Zane Ma Rhea & Lynette Russell, 'Introduction: Understanding Koorie Plant Knowledge Through the Ethnobotanic Lens. A Tribute to Beth Gott', *The Artefact*, 35, 2014, pp. 3–9.
23 Allen, p. 312.
24 'Riparian' means growing on or near water.
25 Zola, Gott & Koorie Heritage Trust, p. 9.
26 Thomas Mitchell, 'August 10, 1832' entry in *Journal of an Expedition into the Interior of Tropical Australia, in Search of a Route from Sydney to the Gulf

of Carpentaria, Longman, Brown, Green and Longmans, London, 1848, <gutenberg.net.au/ebooks/e00035.html>.
27 Local Land Services Western Region, *Ecological Cultural Knowledge – Paakantyi (Barkindji)*.
28 Personal communication with Uncle Badger Bates on the Barka at Menindee, April 2021.
29 Zola, Gott & Koorie Heritage Trust, p. 9.
30 Uncle Badger Bates & Aunty Sarah Martin, *Following Granny Moysey: Kurnu Paakantyi Stories from the Darling, Warrego and Paroo Rivers* [audio recording], 2010. Permission for use granted by Uncle Badger Bates and Dr Aunty Sarah Martin.
31 Anthony is a Ngiyampaa artist and an expert carver who, like David and Uncle Badger Bates, lives and works in Broken Hill, New South Wales. Uncle Badger has supported and mentored these younger men in their carving practice, having been taught to carve as a child by his Granny Moysey (c. 1880–1976), known and widely revered as a *miikitya nhuungku* – clever woman (Bates & Martin, *Following Granny Moysey*).
32 Personal communication with David Doyle, 23 October 2021.
33 Clarke, p. 32.
34 Clarke, p. 33.
35 Bates & Martin, *Following Granny Moysey*
36 Beckett & Hercus, p. 2.
37 Local Land Services Western Region, *Ecological Cultural Knowledge – Paakantyi (Barkindji)*.
38 Allen (ed.), *Australia: William Blandowski's Illustrated Encyclopaedia of Aboriginal Australia*, p. 70.
39 Clarke, p.108.
40 Personal communication with Uncle Badger Bates, 8 November 2021.
41 Bates & Martin, *Following Granny Moysey*; see also Local Land Services Western Region, *Ecological Cultural Knowledge – Barkindji (North of Pooncarie): Knowledge Shared by the Barkindji People*, Local Land Services Western Region, 2016, p. 18, <lls.nsw.gov.au/__data/assets/pdf_file/0011/737624/Barkindji_Booklet_WEB-updated.pdf>; and Clarke.
42 Local Land Services Western Region, *Ecological Cultural Knowledge – Paakantyi (Barkindji)*.
43 Local Land Services Western Region, *Ecological Cultural Knowledge – Barkindji (North of Pooncarie)*, p. 18.
44 Bonney, p. 132.

NOTES

45 Bonney, p. 128.
46 Local Land Services Western Region, *Ecological Cultural Knowledge – Barkindji (North of Pooncarie)*.
47 Zola, Gott & Koorie Heritage Trust, p. 6.
48 Personal communication with Uncle Badger Bates, 8 November 2021.
49 Zola, Gott & Koorie Heritage Trust; see also Fred Cahir, 'Murnong: Much More than a Food', *The Artefact*, 35(1), 2012, pp. 29–39 (p. 33), <search.informit.org/doi/10.3316/informit.362998729686860>.
50 Ronald M Berndt & T Harvey Johnston, 'Death, Burial, and Associated Ritual at Ooldea, South Australia', *Oceania*, 12(3), 1942, pp. 189–208 (p. 208), <jstor.org/stable/40327948>.
51 I call Dr Sarah Martin 'Aunty' because she is married to Uncle Badger and because I respect her as a knowledge holder and as an elder. She has been so generous to me in sharing the knowledge she has amassed through her work with our community over many decades.
52 Dr Aunty Sarah Martin, 'Palaeoecological Evidence Associated with Earth Mounds of the Murray Riverine Plain, South-Eastern Australia', *Environmental Archaeology*, 16(2), 2011, pp. 162–72, <doi.org/10.1179/174963111X13110803261056>.
53 Jessyca Hutchens, 'Blak Yarn' for *Emu Sky* [exhibition], Old Quad, University of Melbourne, November 2021–July 2022, <emusky.culturalcommons.edu.au/exhibition/>.https://emusky.culturalcommons.edu.au/blak_yarn/in-abundance/>.
54 Grace Taylor, 'Branching Out: Making Graphene from Gum Trees', RMIT University, June 2018, <rmit.edu.au/news/all-news/2019/jun/graphene-from-gum-trees>.

5. CUMBUNGI

1 This chapter draws on work discussed in Chapter 6 of my book *Hope and Grief in the Anthropocene*, Routledge, Abingdon, 2016.
2 Unless otherwise noted, the historical references in this chapter are taken from Beth Gott, 'Cumbungi, *Typha* Species: A Staple Aboriginal Food in Southern Australia', *Australian Aboriginal Studies*, 1, 1999, pp. 33–50, where detailed sources can be found. The outstanding contributions of ethnobotanist Dr Beth Gott were also honoured in a special issue of *The Artefact* in 2012.
3 Sylvia J Hallam, 'Aboriginal Women as Providers: The 1830s on the Swan', *Aboriginal History*, 15, 1991, pp. 38–53.

NOTES

4   Paul Irish, *Hidden in Plain View: The Aboriginal People of Coastal Sydney*, New South Publishing, Sydney, 2017.
5   Names discussed by Gott and Hallam, and in Greg Keighery & Steve Mccabe, 'Status of *Typha orientalis* in Western Australia', *Western Australian Naturalist*, 3, 2015, pp. 30–5.
6   Daisy Bates 1938, pp. 69–70, cited in Hallam, p. 38.
7   Oliver Morton, *Eating the Sun: How Plants Power the Planet*, Fourth Estate, London, 2007.
8   Anna Revedin et al., 'Thirty Thousand-Year-Old Evidence of Plant Food Processing', *Proceedings of the National Academy of Sciences*, 107(44), 2010, pp. 18815–19.
9   Karen Hardy, Jennie Brand-Miller, Katherine D Brown, Mark G Thomas & Les Copeland, 'The Importance of Dietary Carbohydrate in Human Evolution', *The Quarterly Review of Biology*, 90(3), 2015, pp. 251–68.
10 Julia F Morton, 'Cattails (*Typha* spp.): Weed Problem or Potential Crop?', *Economic Botany*, 29(1), 1975, pp. 7–29.
11 Carl Linnaeus, *Species Plantarum*, Laurentius Salvius, Stockholm, 1753.
12 M Finlayson, RI Forrester, DS Mitchell & AJ Chick, 'Identification of Native *Typha* Species in Australia', *Australian Journal of Botany*, 33, 1985, pp. 101–7; see also Gott, 1999.
13 Kim Changkyun & Choi Hong-Keun, 'Molecular Systematics and Character Evolution of *Typha* (Typhaceae) Inferred from Nuclear and Plastid DNA Sequence Data', *Taxon*, 60(5), 2011, pp. 1417–28.
14 Chen Peidong, Cao Yudan, Bao Beihua, Zhang Li & Ding Anwei, 'Antioxidant Capacity of *Typha Angustifolia* Extracts and Two Active Flavonoids', *Pharmaceutical Biology*, 55(1), 2017, pp. 1283–8.
15 Jane Roberts & Heidi Kleinert (eds), *Managing Typha and Phragmites: Report from Workshop Held 16th June 2014*, North Central Catchment Management Authority, Victoria, 2015.
16 *Cumbungi Fact Sheet*, Queensland Dept of Agriculture and Fisheries, 2017.
17 J Hale & David L Morgan, *Ecological Character Description for the Lakes Argyle and Kununurra Ramsar Site*, Dept of Sustainability, Environment, Water, Population and Communities, Canberra, 2010.
18 Diego Bonetto, 'How to Cook Nettles', *Bundanon Siteworks – The Blog*, 2010, <siteworksblog.wordpress.com/2010/9/14/on-how-to-cook-nettles/>.
19 LM Mitich, 'Common Cattail, *Typha latifolia* L.', *Weed Technology*, 14(2), 2000, pp. 446–50.
20 Gott, p. 41.

21 Dilek Demirezen & Ahmet Aksoy, 'Accumulation of Heavy Metals in *Typha angustifolia* (L.) and *Potamogeton pectinatus* (L.) Living in Sultan Marsh (Kayseri, Turkey)', *Chemosphere*, 56(7), 2004, pp. 685–96.

22 Woranan Nakbanpote, Orapan Meesungnoen & Majeti Narasimha Vara Prasad, 'Potential of Ornamental Plants for Phytoremediation of Heavy Metals and Income Generation', in Majeti Narasimha Vara Prasad (ed.), *Bioremediation and Bioeconomy*, Elsevier, Amsterdam, 2016, pp. 179–217.

23 AA Adams, A Raman, DS Hodgkins & HI Nicol, 'Accumulation of Heavy Metals by Naturally Colonising *Typha domingensis* (Poales: Typhaceae) in Waste-Rock Dump Leachate Storage Ponds in a Gold–Copper Mine in the Central Tablelands of New South Wales, Australia', *International Journal of Mining, Reclamation and Environment*, 27(4), 2013, pp. 294–307.

24 Martin Krus, Werner Theuerkorn, Theo Großkinsky & Hartwig Künzel, 'New Sustainable and Insulating Building Material Made of Cattail', *Notulae Scientia Biologicae*, 2014, pp. 1252–60.

25 Younouss Dieye et al., 'Thermo-Mechanical Characterization of a Building Material Based on *Typha Australis*', *Journal of Building Engineering*, 9, 2017, pp. 142–6.

6. YAMS

1 This chapter draws heavily on and synthesises work previously published with Jenny Atchison and Richard Fullagar, in Lesley Head, Jennifer Atchison & Richard Fullagar, 'Country and Garden: Ethnobotany, Archaeobotany and Aboriginal Landscapes near the Keep River, Northwestern Australia', *Journal of Social Archaeology*, 2, 2002, pp. 173–96; Jennifer Atchison & Lesley Head, 'Yam Landscapes: The Biogeography and Social Life of Australian *Dioscorea*', *The Artefact*, 35, 2012, pp. 59–74; Jennifer Atchison & Lesley Head, 'Exploring Human-Plant Entanglements: The Case of Australian *Dioscorea* Yams', in David Frankel, Jennifer Webb & Susan Lawrence (eds), *Archaeology in Environment and Technology: Intersections and Transformations*, Routledge, London, 2013, pp. 167–80. In turn, those papers review and reference many historical and ethnographic examples, recorded by dozens of people. Please consult those papers for detailed sourcing.

2 Atchison & Head, 'Exploring Human-Plant Entanglements', p. 176.

3 Head, Atchison & Fullagar, p. 179.

4 Head, Atchison & Fullagar, p. 179.

5 Head, Atchison & Fullagar, p. 180.

6 Head, Atchison & Fullagar, p. 181.

7   Head, Atchison & Fullagar, p. 180.
8   Head, Atchison & Fullagar, p. 180.
9   Sylvia J Hallam, 'Yams, Alluvium and Villages on the Western Coastal Plain', *54th Congress of the Australian and New Zealand Association for the Advancement of Science (ANZAAS)*, Australian Institute of Aboriginal Studies, Canberra, 1986.
10 Sylvia J Hallam, 'Plant Usage and Management in Southwest Australian Aboriginal Societies', in David R Harris & Gordon C Hillman (eds), *Foraging and Farming: The Evolution of Plant Exploitation*, Routledge, London, [1989] 2016, pp. 136–51.
11 Peter Veth, Cecilia Myers, Pauline Heaney & Sven Ouzman, 'Plants Before Farming: The Deep History of Plant-Use and Representation in the Rock Art of Australia's Kimberley Region', *Quaternary International*, 489, 2018, pp. 26–45; Sven Ouzman, Peter Veth, Cecilia Myers, Pauline Heaney & Kevin Kenneally, 'Plants Before Animals?: Aboriginal Rock Art as Evidence of Ecoscaping in Australia's Kimberley', in Bruno David & Ian J McNiven (eds), *The Oxford Handbook of the Archaeology and Anthropology of Rock Art*, 2017, <doi.org/10.1093/oxfordhb/9780190607357.013.31>.
12 Judith W Hammond, 'The Spatial Distribution of Unique Motifs Featuring *Dioscorea bulbifera*, the Round Yam, in Western Arnhem Land Rock Art', *Australian Archaeology*, 85(2) 2019, <doi.org/10.1080/03122417.2019.1686237>.
13 Jane C Goodale, *Tiwi Wives: A Study of the Women of Melville Island, North Australia*, University of Washington Press, Seattle, 1971.
14 These results were recorded in three separate studies made across some fifty years; see Frederick D McCarthy & M McArthur, 'The Food Quest and the Time Factor in Aboriginal Economic Life', in Charles P Mountford (ed.), *Records of the American-Australian Scientific Expedition to Arnhem Land*, Melbourne University Press, Melbourne, 1960, pp. 145–94; Rhys Jones & Betty Meehan, 'Plant Foods of the Gidjingali: Ethnographic and Archaeological Perspectives from Northern Australia on Tuber and Seed Exploitation', in Harris & Hillman (eds), *Foraging and Farming: The Evolution of Plant Exploitation*, pp. 120–51; and Julian Gorman, Glenn Wightman, Peter Whitehead & Jon Altman, 'Case Study: Long Yam. Feasibility of Small Scale Commercial Native Plant Harvests by Indigenous Communities', in Peter Whitehead et al. (eds), *Rural Industries Research and Development Corporation (RIRDC) Land & Water Australia Joint Venture Agroforestry Program*, Rural Industries Research and Development Corporation, Canberra, 2006, pp. 143–50.

NOTES

15 George Grey, *Journals of Two Expeditions of Discovery in North-West and Western Australia, During the Years 1837–1839*, Vol. 2, T & W Boone, London, [1841] 1964, p. 293.

16 Ian M Crawford, 'Traditional Aboriginal Plant Resources in the Kalumburu Area: Aspects in Ethno-Economics' [unpublished], *Records of the Western Australian Museum Supplement No. 15*, Western Australian Museum, Perth, 1982.

17 Hammond, p. 4.

18 Grey; Citations of George Grey's work are taken from papers by archaeologist Sylvia Hallam (see notes 9 and 10 of this chapter), who analysed them in detail.

19 Grey, quoted in Hallam, 'Yams, Alluvium and Villages on the Western Coastal Plain', p. 118.

20 Grey, quoted in Hallam, pp. 119–20.

21 Grey, quoted in Hallam, 'Plant Usage and Management in Southwest Australian Aboriginal Societies', p. 142.

22 Hallam, 'Yams, Alluvium and Villages on the Western Coastal Plain'; Hallam, 'Plant Usage and Management in Southwest Australian Aboriginal Societies'.

23 R Jones, 'The Neolithic, Palaeolithic and the Hunting Gardeners: Man and Land in the Antipodes', in RP Suggate & MM Cresswell (eds), *9th INQUA Congress*, The Royal Society of New Zealand, Wellington, 1973; see also Rosemary Hill & Adelaide Baird, 'Kuku-Yalanji Rainforest Aboriginal People and Carbohydrate Resource Management in the Wet Tropics of Queensland, Australia', *Human Ecology*, 31(1), 2003, pp. 27–51.

24 Jeremy Russell-Smith et al., 'Aboriginal Resource Utilisation and Fire Management Practice in Western Arnhem Land, Monsoonal Northern Australia: Notes for Prehistory, Lessons for the Future', *Human Ecology*, 25(2), 1997, pp. 159–95.

25 India E Dilkes-Hall, Susan O'Connor & Jane Balme, 'People-Plant Interaction and Economic Botany over 47,000 Years of Occupation at Carpenter's Gap 1, South Central Kimberley', *Australian Archaeology*, 85(1), 2019, pp. 30–47, <doi.org/10.1080/03122417.2019.1595907>; see also India E Dilkes-Hall, Jane Balme, Susan O'Connor & Emilie Dotte-Sarout, 'Archaeobotany of Aboriginal Plant Foods During the Holocene at Riwi, South Central Kimberley, Western Australia', *Vegetation History and Archaeobotany*, 29(3), 2020, pp. 309–25; and Jennifer Atchison, Lesley Head & Richard Fullagar, 'Archaeobotany of Fruit Seed Processing in a Monsoon

Savanna Environment: Evidence from the Keep River Region, Northern Territory, Australia', *Journal of Archaeological Science*, 32(2), 2005, pp. 167–81.
26. Jane C Goodale, 'An Example of Ritual Change among the Tiwi of Melville Island', in Arnold R Pilling & Richard A Waterman (eds), *Diprotodon to Detribalisation*, Michigan State University Press, East Lansing, 1970, pp. 350–66; see also Goodale, *Tiwi Wives*.
27. Atchison & Head, *Yam Landscapes*.
28. Lesley Hughes, Will Steffen, Martin Rice & Alix Pearce, *Feeding a Hungry Nation: Climate Change, Food and Farming in Australia*, Climate Council of Australia, 2015.
29. 'Fareshare Abbotsford: Growing Sweet Potatoes in Victoria', *The Weekly Times*, 14 June 2017, <weeklytimesnow.com.au/agribusiness/farm-magazine/fareshare-abbotsford-growing-sweet-potatoes-in-victoria/news-story/04e0111db6477bf638e6f2036298de08>.

7. SPINIFEX

1. Includes plants previously named *Plectrachne*.
2. Harshi K Gamage et al., 'Indigenous and Modern Biomaterials Derived from *Triodia* ('Spinifex') Grasslands in Australia', *Australian Journal of Botany*, 60(2), 2012, pp. 114–27 (p. 115).
3. Table adapted from Table 2 in Gamage et al. For detailed sources see Heidi T Pitman & Lynley A Wallis, 'The Point of Spinifex: Aboriginal Uses of Spinifex Grasses in Australia', *Ethnobotany Research and Applications*, 10, 2012, pp. 109–31.
4. Powell, Fensham and Memmott also document some interesting museum examples from Queensland, see Owen Powell, Roderick J Fensham & Paul Memmot, 'Indigenous Use of Spinifex Resin for Hafting in North-Eastern Australia', *Economic Botany*, 67(3), 2013, pp. 210–24.
5. Kim Akerman, 'Appendix: *Triodia* Resin Manufacturing Techniques', in Veerle Rots et al., 'Hafted Tool-Use Experiments with Australian Aboriginal Plant Adhesives: *Triodia* Spinifex, *Xanthorrhoea* Grass Tree and *Lechenaultia divaricata Mindrie*', *EXARC Journal*, 2020/1, <exarc.net/ark:/88735/10487>.
6. Pitman & Wallis, p. 112.
7. Akerman, p. 7.
8. P Green, 'Sticky Harvest: From Harvest to Haft', n.d., unpublished manuscript.
9. Rots et al.
10. Summarised in Gamage et al.

NOTES

11 India E Dilkes-Hall, Susan O'Connor & Jane Balme, 'People-Plant Interaction and Economic Botany over 47,000 Years of Occupation at Carpenter's Gap 1, South Central Kimberley', *Australian Archaeology*, 85(1), 2019, pp. 30–47, <doi.org/10.1080/03122417.2019.1595907>.

12 Mike Smith, 'Late Quaternary Landscapes in Central Australia: Sedimentary History and Palaeoecology of Puritjarra Rock Shelter', *Journal of Quaternary Science*, 24(7), 2009, pp. 747–60.

13 Richard Fullagar & Judith Field, 'Pleistocene Seed-Grinding Implements from the Australian Arid Zone', *Antiquity*, 71(272), 1997, pp. 300–7; see also Richard Fullagar, Judith Field & Lisa Kealhofer, 'Grinding Stones and Seeds of Change: Starch and Phytoliths as Evidence of Plant Food Processing', in Yorke M Rowan and Jennie R Ebeling (eds), *New Approaches to Old Stones: Recent Studies of Ground Stone Artifacts*, Equinox, London, 2008, pp. 159–72.

14 Rots et al.

15 Rots et al., p. 5.

16 Rots et al., p. 8.

17 Elspeth Hayes, Richard Fullagar, Ken Mulvaney & Kate Connell, 'Food or Fibercraft? Grinding Stones and Aboriginal use of *Triodia* Grass (Spinifex)', *Quaternary International*, 468, 2018, pp. 271–83.

18 Pitman & Wallis.

19 Peter Bindon, *Useful Bush Plants*, Western Australian Museum, Perth, 1996, p. 257.

20 Pitman & Wallis, p. 120.

21 Hayes et al., p. 278.

22 Mickey Dewar, 'Terry, Michael (1899–1981)', *Australian Dictionary of Biography*, 2012, <adb.anu.edu.au/biography/terry-michael-15670>.

23 'Kylie' is a Nyungar word for boomerang.

24 Michael Terry, *Through a Land of Promise: With Gun, Car and Camera in the Heart of Northern Australia*, Herbert Jenkins, London, 1927, p. 265.

25 Terry, p. 266.

26 Hayes et al., p. 280.

27 Hayes et al., p. 279.

28 Hayes et al., p. 272.

29 Heidi T Pitman, 'A Cake of Spinifex Resin', in Steve Brown, Anne Clarke & Ursula Frederick (eds), *Object Stories*, Routledge, New York, 2015, pp. 93–101.

30 Pitman, p. 2.

31 Pitman, p. 3.

32 Pitman, p. 4.

33 Steven Hemming, 'Objects and Specimens: Conservative Politics and the SA Museum's Aboriginal Cultures Gallery', *Overland*, 171, 2003, pp. 64–9 (p. 64).
34 Pitman, p. 8.
35 Paul Memmott, Richard Hyde & Timothy O'Rourke, 'Biomimetic Theory and Building Technology: Use of Aboriginal and Scientific Knowledge of Spinifex Grass', *Architectural Science Review*, 52(2), 2009, pp. 117–25.
36 Terri Janke, 'From Smokebush to Spinifex: Towards Recognition of Indigenous Knowledge in the Commercialisation of Plants', *International Journal of Rural Law and Policy*, 1, 2018, pp. 1–37, <doi.org/10.5130/ijrlp.i1.2018.5713>.
37 Janke, p. 5.
38 Gamage et al.
39 Gamage et al., p. 123.

8. QUANDONGS

1 Neville Bonney, *Jewel of the Australian Desert, Native Peach (Quandong): The Tree with the Round Red Fruit*, self-published by Neville Bonney, 2013, p. 17.
2 Sandy Laurie, *Australian Native Foods and Botanicals – 2019/20 Market Study*, ANFAB, 2020, <anfab.org.au/edit/research_projects/ANFAB_2020_Market%20Study.pdf>.
3 Bonney, p. 9.
4 Maurizio Rossetto et al., 'From Songlines to Genomes: Prehistoric Assisted Migration of a Rain Forest Tree by Australian Aboriginal People', *PLOS ONE*, 12(11), 2017, <doi.org/10.1371/journal.pone.0186663>.
5 Richard A Gould, 'Subsistence Behaviour among the Western Desert Aborigines in Australia', *Oceania*, 39(4), 1969, pp. 253–74; see also N Peterson, 'The Traditional Pattern of Subsistence to 1975', in Basil S Hetzel & HJ Frith (eds), *The Nutrition of Aborigines in Relation to the Ecosystem of Central Australia: Papers Presented at a Symposium, CSIRO, 23–26 October 1976, Canberra*, CSIRO, Melbourne, 1977, pp. 25–35; and Cliff Goddard & Arpad Kalotas (comps, eds), *Punu: Yankunytjatjara Plant Use*, IAD Press, Alice Springs, 1985; and Yasmina Sultanbawa & Fazal Sultanbawa (eds), *Australian Native Plants: Cultivation and Uses in the Health and Food Industries*, CRC Press, Boca Raton, Florida, 2016, p. 150, <doi.org/10.1201/b20635>; and Jospeh H Maiden, *The Useful Native Plants of Australia*, Turner and Henderson, Sydney, 1889, cited in Colin Pardoe, Richard Fullagar & Elspeth

Hayes, 'Quandong Stones: A Specialised Australian Nut-Cracking Tool', *PLOS ONE*, 14(10), 2019, <doi.org/10.1371/journal.pone.0222680>. This chapter draws heavily on Pardoe, Fullagar & Hayes and the references therein.
6   Pardoe, Fullagar & Hayes, p. 3.
7   Goddard & Kalotas, p. 35, cited in Pardoe, Fullagar & Hayes, p. 5.
8   Goddard & Kalotas, p. 37, cited in Pardoe, Fullagar & Hayes, p. 2.
9   Pardoe, Fullagar & Hayes, p. 5.
10 Nelly Zola, Beth Gott & Koorie Heritage Trust, Koorie Plants, *Koorie People: Traditional Aboriginal Food, Fibre and Healing Plants of Victoria*, Koorie Heritage Trust, Melbourne, 1992, p. 29.
11 Zola, Gott & Koorie Heritage Trust, p. 29.
12 Jennifer Isaacs, *Bush Food: Aboriginal Food and Herbal Medicine*, Lansdowne, Sydney, 1987, p. 239.
13 Colin Pardoe & Dan Hutton, 'Aboriginal Heritage as Ecological Proxy in South-Eastern Australia: A Barapa Wetland Village', *Australasian Journal of Environmental Management*, 28, 2020, pp. 17–13, <doi.org/10.1080/14486563.2020.1821400>.
14 Pardoe, Fullagar & Hayes, pp. 4–5.
15 Heath Dunstan, Singarayer K Florentine, Maria Calviño-Cancela, Martin E Westbrooke & Grant C Palmer, 'Dietary Characteristics of Emus (*Dromaius novaehollandiae*) in Semi-Arid New South Wales, Australia, and Dispersal and Germination of Ingested Seeds', *Emu*, 113(2), 2013, pp. 168–76.
16 Margaret Raven, Daniel Robinson & John Hunter, 'The Emu: More-Than-Human and More-Than-Animal Geographies', *Antipode*, 53, 2021, pp. 1526–45, <doi.org/10.1111/anti.12736>.
17 The implements are held by Local Aboriginal Councils and other Aboriginal groups in local museums and on pastoral properties.
18 Pardoe, Fullagar & Hayes.
19 Pardoe, Fullagar & Hayes, p. 6.
20 R Clegg quoted in Pardoe, Fullagar & Hayes, p. 8.
21 Pardoe, Fullagar & Hayes, p. 7.
22 Pardoe, Fullagar & Hayes, p. 22.
23 Pardoe, Fullagar & Hayes, pp. 25–6.
24 Pardoe, Fullagar & Hayes, p. 26.
25 Alison Downing, Brian Atwell, Karen Marais, David Edgecome & Kevin Downing, *Santalum, Sandalwoods and Quandongs*, Dept of Biological Sciences Macquarie University, nd, <bio.mq.edu.au/wp-content/uploads/2021/05/Plant-of-the-week-Santalum-sandalwood-and-quandongs.pdf>.

26 Joanna Prendergast, 'Western Australia's $40m Native Sandalwood Industry Risks Collapse, Industry Groups Warn', *ABC News*, 1 March 2021, <abc.net.au/news/rural/2021-03-01/native-sandalwood-groups-call-for-change-risks-collapse/13197368>.
27 Eden Hynninen, 'Ancient Tree Species, Northern Sandalwood, Brought Back from Brink of Extinction in Victoria', *ABC News*, 13 May 2021, <abc.net.au/news/2021-05-13/tree-species-on-the-brink-of-extinction-nurtured-in-victoria/100134740>.

9. RESPECTING KNOWLEDGE

1 Kim Honan, 'Demand for Bush Food is Booming, So Why Are So Few Indigenous People Involved in the Sector', *ABC News*, 9 July 2021, <abc.net.au/news/2021-07-09/native-food-sector-seeks-connection-with-indigenous-australia/100271318>.
2 Emma Woodward, Diane Jarvis & Kirsten Maclean, *The Traditional Owner-Led Bush Products Sector: An Overview*, CSIRO, Canberra, 2019, p. 235.
3 Yasmina Sultanbawa & Fazal Sultanbawa (eds), *Australian Native Plants: Cultivation and Uses in the Health and Food Industries*, CRC Press, Boca Raton, Florida, 2016, p. 150, <doi.org/10.1201/b20635>.
4 Sandy Laurie, *Australian Native Foods and Botanicals – 2019/20 Market Study*, ANFAB, 2020, p. 6, <anfab.org.au/edit/research_projects/ANFAB_2020_Market%20Study.pdf>.
5 Woodward, Jarvis & Maclean, p. 233.
6 Graham Dutfield, 'Traditional Knowledge, Intellectual Property and Pharmaceutical Innovation: What's Left to Discuss?', in Matthew David & Debora Halbert (eds), *The SAGE Handbook of Intellectual Property*, SAGE Publications Ltd., London, 2015, p. 651.
7 Diane Jarvis et al., 'The Learning Generated Through Indigenous Natural Resources Management Programs Increases Quality of Life for Indigenous People – Improving Numerous Contributors to Wellbeing', *Ecological Economics*, 180, 2021, <doi.org/10.1016/j.ecolecon.2020.106899>.
8 See <naakpa.com.au/about-naakpa>.
9 See *The Victorian Traditional Owner Native Foods and Botanicals Strategy*, Federation of Victorian Traditional Owner Corporations, n.d., <fvtoc.com.au/native-foods-and-botanicals>.
10 See <fnbbaa.com.au/about>.
11 Terri Janke, Maiko Sentina & Terri Janke and Company, *Indigenous Knowledge: Issues for Protection and Management*, IP Australia, 2018,

<ipaustralia.gov.au/sites/default/files/ipaust_ikdiscussionpaper_28march 2018.pdf>.

12 Daniel Robinson & Margaret Raven, 'Identifying and Preventing Biopiracy in Australia: Patent Landscapes and Legal Geographies for Plants with Indigenous Australian Uses', *Australian Geographer*, 48(3), 2017, pp. 311–31, <doi.org/10.1080/00049182.2016.1229240>.

13 Terri Janke et al., *Comparative Study on Genetic Resources, Traditional Knowledge and Traditional Cultural Expressions (GRTKTCE)*, AANZFTA, p. 33, <aanzfta.asean.org/uploads/2021/10/Comparative-Study-GRTKTCE_Final-For-Public.pdf>.

14 Robinson & Raven, pp. 311–31.

15 *The Legal Regulation of Biodiscovery in Australia*, Uniquely Australian Foods, 2020, <uniquelyaustralianfoods.org/wp-content/uploads/2020/03/UAF16-The-Legal-Regulation-of-Biodiscovery-in-Australia-Fact-Sheet-0721.pdf>.

16 Laurie, *2019/20 Market Study*, p. 4.

17 Alana Mann, *Food in a Changing Climate*, Emerald Publishing Limited, Bingley, 2021, pp. 72–7.

18 'Good for the Economy', Australian Almonds, n.d., <australianalmonds.com.au/sustainable-almonds/?v=6cc98ba2045f>.

19 Jacquelyn Simpson, *Almond Industry Expansion*, Dept of Primary Industries NSW, July 2016, <dpi.nsw.gov.au/__data/assets/pdf_file/0004/586435/almond-industry-expansion.pdf>; see also Cara Jeffery, Jessica Schremmer & Michael Condon, 'Almond Milk Might Not Be as "Planet-Friendly" as You Think, Expert Sparks Fiery Debate', *ABC News*, 24 February 2021, <abc.net.au/news/rural/2021-02-24/almond-milk-production-and-water-use-fuel-sustainability-debate/13186968>.

20 Annette McGivney, '"Like Sending Bees to War": The Deadly Truth Behind Your Almond Milk Obsession', *The Guardian*, 8 January 2020, <theguardian.com/environment/2020/jan/07/honeybees-deaths-almonds-hives-aoe>.

21 Mann, pp. 72–7.

22 Lorena Allam, 'Australian Researchers Find Native Grasses Could Be Grown for Mass Consumption', *The Guardian*, 10 November 2020, <theguardian.com/australia-news/2020/nov/10/australian-researchers-find-native-grasses-could-be-grown-for-human-consumption#:~:text=Mulga%2C%20Mitchell%20grass%20and%20Kangaroo,these%20benefits%20need%20further%20study>; see also Nicholas A Moore, James C Camac & John W Morgan, 'Effects of Drought and Fire on Resprouting Capacity of 52 Temperate Australian Perennial Native

Grasses', *New Phytologist*, 221(3), 2019, pp. 1424–33, <doi.org/10.1111/nph.15480>.
23 Desert Support Services, 'Integrated Buffel Grass Management Plan for the Great Victoria Desert', Rural Solutions SA, Perth, 2018.
24 Lorena Allam & Isabelle Moore, '"Time to Embrace History of Country": Bruce Pascoe and the First Dancing Grass Harvest in 200 Years', *The Guardian*, 13 May 2020, <theguardian.com/artanddesign/2020/may/13/its-time-to-embrace-the-history-of-the-country-first-harvest-of-dancing-grass-in-200-years>.
25 Megan Ferguson et al., 'Traditional Food Availability and Consumption in Remote Aboriginal Communities in the Northern Territory, Australia', *Australian and New Zealand Journal of Public Health*, 41(3), 2017, pp. 294–8, <doi.org/10.1111/1753-6405.12664>; see also Jennifer Browne, Mark Lock, Troy Walker, Mikaela Egan & Kathryn Backholer, 'Effects of Food Policy Actions on Indigenous Peoples' Nutrition-Related Outcomes: A Systematic Review', *BMJ Global Health*, 5(8), 2020, <doi.org/10.1136/bmjgh-2020-002442>.
26 Fred Cahir, 'Murnong: Much More than a Food', *The Artefact*, 35(1), 2012, pp. 29–39, <search.informit.org/doi/10.3316/informit.362998729686860>.
27 Christopher P Burgess et al., 'Healthy Country, Healthy People: The Relationship Between Indigenous Health Status and "Caring for Country"', *The Medical Journal of Australia*, 190(10), 2009, pp. 567–72, <doi.org/10.5694/j.1326-5377.2009.tb02566.x>; see also Jarvis et al., 'The Learning Generated Through Indigenous Natural Resources Management Programs Increases Quality of Life for Indigenous People'.
28 Rosalie Schultz, Tammy Abbott, Jessica Yamaguchi & Sheree Cairney, 'Australian Indigenous Land Management, Ecological Knowledge and Languages for Conservation', *EcoHealth*, 16(3), 2019, pp. 171–6, <doi.org/10.1007/s10393-018-1380-z>; see also Douglas Bird, Rebecca Bliege Bird, Brian Codding & Nyalangka Taylor, 'A Landscape Architecture of Fire: Cultural Emergence and Ecological Pyrodiversity in Australia's Western Desert', *Current Anthropology*, 57(s13), 2016, <doi.org/10.1086/685763>.
29 Allam; see also Zane Ma Rhea, *Frontiers of Taste: Food Sovereignty, Sustainability, and Indigenous-Settler Relations in Australia*, Springer Nature, Singapore, 2016, pp. 185, 189.
30 Cathy Robinson, Emily Gerrard, Tray May & Kirsten Maclean, 'Australia's Indigenous Carbon Economy: A National Snapshot', *Geographical Research*, 52(2), 2014, pp. 123–32, <doi.org/10.1111/1745-5871.12049>.

10. FUTURES

1 Zena Cumpston, 'Illuminating Indigenous Culture Through Plants', *Pursuit*, n.d., <pursuit.unimelb.edu.au/articles/illuminating-indigenous-culture-through-plants>.
2 Antasia Azure, 'Owning the Future: An Interview with Melissa George', *Cultural Survival Quaterly*, 29(2), 2005, <culturalsurvival.org/publications/cultural-survival-quarterly/owning-future-interview-melissa-george>.
3 Fred Cahir, 'Murnong: Much More than a Food', *The Artefact*, 35(1), 2012, pp. 29–39, <search.informit.org/doi/10.3316/informit.362998729686860>.
4 Zane Ma Rhea, *Frontiers of Taste: Food Sovereignty, Sustainability, and Indigenous-Settler Relations in Australia*, Springer Nature, Singapore, 2016, pp. 184, 188–9, 205–6.
5 'Funding Certainty for Indigenous Rangers' [media release], Dept of the Prime Minister and Cabinet, 10 March 2020, <ministers.pmc.gov.au/wyatt/2020/funding-certainty-indigenous-rangers>.
6 *The Victorian Traditional Owner Native Foods and Botanicals Strategy*, Federation of Victorian Traditional Owner Corporations, n.d., <fvtoc.com.au/native-foods-and-botanicals>.
7 Christopher Mayes, *Unsettling Food Politics: Agriculture, Dispossession and Sovereignty in Australia*, Rowman & Littlefield International, London, 2018, pp. 9, 11, 167, 205, 206, 209.

INDEX

Note: Page references in **bold** refer to images, figures or captions.

Ackerman, Kim, 133
Allen, Harry, 82, 83
Angas, George, 101–2
Atchison, Jenny, 26–7, 28–9, 115, 128
Atkinson, Henry, 159
Australia's First Naturalists (Russell and Olsen), 36

backyards
   and settler-colonial relations to plants, 27–8
Balbuk, Fanny, 103–4, 112
Balme, Jane, 136
Bates, Uncle Badger, 10, **32**, 84, 85–6, 87–8, 90, 94, 99
Bates, Daisy
   The Passing of the Aborigines, 104
Beveridge, Peter, 101, 103
Blandowski, William, 38, 82, 90
   recorded NjeriNjeri names of animals, fish, insects, 39
Bolin Bolin. *See* Melbourne (Narrm)
Bonetto, Diego, 109
Bonney, Edward, 71
Bonney, Frederic, 71, 72, 78, 92
   and photographs of Aboriginal people, 69–72, 73, 75
Bromwell, Tuesday, 159

clans/language groups
   Barkandji (also Barkindji or Paakantyi), 9, 69, **70**, 72, 73, 74–5, 85, 86, 88, 92, (sharing knowledge), 91; art, 89; and Barka/Darling River, 10, 69; Country, 10, 69, **70**, 78, 81, 82, 84, 91, 94, 153, 154, 156, (and drought), 78, 85; settler encroachment, 71
   Boon Wurrung/Bunurong, 51; Country, 59–60, 64
   of Central Australia, 152
   DjaDja Wurrung: corporation, 46–7
   Gajerrong-Djarradjarrany, 29
   Gunditjmara, 39; Country, 62
   Jaminjung, 25, 115
   Kamilaroi, 189 n 17
   Kaurna, 48
   Kulin Nations, 18
   Madi Madi, 153
   Mirarr: Country, 77
   Murrinh-patha, 25, 115
   Ngiyampaa, 190 n 31
   NjeriNjeri, 38, 39
   Nyungar, 103
   palawa, 21
   Tati Tati, 30
   Tiwi, 122, 127, 128, 129
   Wiradjuri, 4, 17, 107, 154, 157, 189 n 15
   Woiwurrung, 17
   Wurundjeri, 48; and Country, 51–6, 67
   WurundjeriWoiwurrung (community), 11–12, 34; and plant knowledge, 11

INDEX

Clean Air Urban Landscapes Hub, 11
  range of researchers, 11
Clegg, Robert, 157
climate change
  and action, 28; in urban areas, 47
  and food security, 129
  and Indigenous knowledges, 7, 47–8
  and Murray-Darling Basin, 110
colonisation, 15
  collectors: and Indigenous
    knowledge, 38, 101
  colonialism, 26; erasing Indigenous
    science, knowledge, 40, 42, 43, 46,
    127; narratives, 36, 37, 44–5
  colonial/European views of
    Indigenous people, 37, 42, 43,
    65–6, 98; agency denied, 62–3,
    64; and child removal policies, 44,
    62; drawings, paintings, accounts
    of, 38, 39, 95; food of, 161–2; as
    'inferior', a 'dying race', 44, 60,
    62; and lack of recognition of
    design, management, 3–4, 51, 60,
    61–2, 175; and myth of 'aimless
    wanderers', 42, 67; relations, 37, 38,
    (power imbalance), 38, 46, 65
  and Country, 15, 64, 65, 67, 73,
    145, 170–1, 177; unsustainable
    practices, 6, 46, 174, 175
  frontier violence, 73
  impact on Aboriginal people, 26, 37,
    99, 147–8, 176; land management,
    ownership denied, 3–4, 25, 27, 125;
    limited access to land, resources,
    25, 163, 170–1, 176, 177
  impressions of landscape, 59;
    lack of an Indigenous voice,
    63; overlooking Indigenous
    management, 60

and racism, 175
terra nullius, 4, 65; amounts to theft,
  60–1, 62, 68
and truth-telling, 175, 177
see also Country
Connell, Kate, 139
Country
  abundance, 74, 97–8, 162, 173
  adaptation over time, 34
  built structures, 125; European
    influences, 90; as extensions of
    Country, 6; houses, 80, 92; shelters/
    dwellings, 2, 29, **70**, 81, 90, 94, 97,
    125, 147; villages, 61, 125, 154
  caring for, maintaining Country, 1, 4,
    12, 29, 34, 55, 89, 175; empowering
    Indigenous communities, 175,
    177; and healing, 12, 51, 175; and
    invasive species, 169–70, (and
    monocultures), 170; knowledge of
    Country, 12, 19, 31, 46, 60–1, 161,
    (shared with Europeans), 37; and
    plants, 50–1, 161
  central to Indigenous people, 2,
    31; as mother, 89, 177; and right
    of access, 74, 170–1; and climate
    change, 14, 47–8, 169
  connection with, 26, 27, 42, 51–3,
    74, 89, 95; ancestral and spiritual
    dimensions, relationships, 31;
    belonging, 32–3, 34, 48; and oral
    history, 38–9
  and degradation, 10, 54; almonds
    and biodiversity, 168; and repair, 49
  and dispossession, 52, 62, 104, 125,
    148, 172; women's resistance,
    104
  fire regimes, 4, 32, 58–9;
    creating new ecosystems, 4, 22,

205

57–8; fire-stick farming, 22; manipulating plants, 4, 20, 24, 25, 32, 125–6; in urban areas, 48; on wetlands, 24, 56; women and men, 25–6

and history, 54; research, 55; and wetlands, 51–6

hunting, fishing on, 26, 87, 89; emus, 90

land management regimes, 4, 6, 13, 15, 20, 21, 22–3, 25, 42, 51, 59–60, 67–8, 82, 160, 173, 176; aquaculture, 61–2; design, 61; and diversity, 57; and flexibility, 61; and fruit tree distribution, 27, 149, 154, 157–8; and respect for, 163; vegetation, 22, 24, 25, 27, 51

and land rights, 23; native title, 29

learning on, 30–1, 74

non-Indigenous misconceptions, 32, 36, 61–2, 64, 67, 161

quarry sites, 119

and stories, 12, 13, 34, 39, 48

and sustainability, 6, 28, 46, 131, 160, 161, 163, 164, 174, 175

traditional foods and health, 170–1; new projects, 171–2; land management, fire, 171

urban areas, 33–4; degradation, 47; reclaiming, 3, 12

see also colonisation; Indigenous culture; Indigenous knowledges; Melbourne (Narrm)

cumbungi (*Typha*, bulrush), 4
for canoes, 106
colonial records, 101–2, 106
cooking: mound ovens, 102–3
cultivation, 24, 109; hard work, 111
and Europeans, 4, 25; now invisible to, 103; seen as a weed, 4, 25, 108–9
fibres, 84–5, 100; and string-making, 101, **101**
as food, 4, 81, 83–4, 100, 109, 111; cakes, bread, 102–3; and Māori, 24
and the future (the Anthropocene), 109–10; as building material, 4, 112, 113; food source, 110, 111; and landscape, 112; tackling pollution, 109–10, 111–12, 113
importance, 81, 83–4, 107
and movement, 106; coloniser, 111; expanded habitat, 108
names for, 107; in Australian languages, 107
and nets, bags, 81, 85, 100, 108
and pollution, 4, 109–10, 111–12
relationships with humans, 105, 108–9, 109–10, 112; and clothing, 108; as medicine, 108; shelter, 108
and starch, 84, 100, 102, 103, 111; and energy, 105
tubers of, 83
and wetlands, 111
widespread, 100, 107; in urban areas, 100
and work of women, 103–4, 105

Darling (Barka/Baaka) River, 10
and greed, mismanagement, 10
Darwin, Charles
theory of evolution, 7, 20, 62
Dilkes-Hall, India, 127, 136
Doughboy (Barkandji woman), **70**, 75, 76, 86, 87, 89, 90, 99
Doyle, David, 74–5, 90, 151, 188 n 2

and Aboriginal culture, 78, 80, 81, 82, 85–6, 88–9, 90, 99, 151
as artist, **155**, 156

environment
challenges, 36; continual growth, 98
and dominion over nature, 97
and Franklin River dam project, 23
Indigenous engagement with, 23
State of the Environment Report 2021, 13
waste and pollution, 97
see also climate change
Everard, Mollie, 152
Everard, Pompey, 152
exhibitions
Emu Sky, 13, 40, 79

feminism, 23–4
Fforde, Cressida, 45
Flannery, Tim
The Future Eaters, 67
Fullagar, Richard, 25, 119, 139, 156

Gammage, Bill, 6
Golson, Jack, 22
Goodale, Jane, 122, 127, 128
Gott, Beth, 82, 104, 107, 111, 153, 191 n 2
Grant, Uncle Stan, 79, 189 n 15
Grey, George, 24, 122, 124–5
Grist, Mark J, 38–9

Hallam, Sylvia, 104, 120, 125–6
Hammond, Judith, 123–4
Hayes, Ebbe, 139, 140, 141, 156
Hayward, Anthony, 87, 190 n 31
Hoddle, Robert, **52**

Horton, David
The Pure State of Nature, 67
Hutchens, Jessyca, 97–8
Hutton, Dan, 154–5

Indigenous communities, 14
and authenticity, 48
cultures, knowledge, 31, 82; and plants, 46, 96
and empowerment: to care for Country, 175
and family histories, 72–3
lack of water rights, 10
and managing Country, 15, 176; and food sovereignty, 176, 177; shared knowledge, 31–2
and seasons, 18
and self-determination, 163, 172; and ICIP rights, 165–6, 167
and Stolen Generations, 17–18, 44
stories, 15
and universities, museums, galleries, 16, 93
in urban environments, 33–4, 48, 170–1; and caring, 34
Indigenous culture
and ancestor beings, 1, 11, 89, 96, 133; Creation stories, 85–6, 89; the Dreaming, 1, 115, 155
art, 79; decoration, 87, 88–9; engraving, 88–9; and information, 88, 120, 123–4; rock art, 120, 121
belief systems, 7
ceremonies, 93, 122, 127; initiation, 92–3; large gatherings, 102
clothing: kangaroo cloaks, 90; possum rugs, 90
and custodial obligations, 12, 48, 74, 89

generational connections, 157
innovation, adaptation, 13, 34, 73
knowledge archive, 6, 20
languages, 189 n 17; and language revival, 7, 38
Law, 89, 177
lifestyle, 73, 75–6, 78
longevity of, 6, 7, 23, 58, 97, 160, 169, 174
non-plant food: animals, fish and birds, 82, 83; hunting, 42
ovens, 94, 102–3; earth mounds as, 96
and smoking tobacco, 90
social systems: gender segregation, 93, 122
Songlines: and Country, 6; as storehouses of knowledge, 2; and travel, 2
stories, 14, 26, 49, 79, 80; oral culture, histories, 72, 188 n 2
trade, exchange, 131, 134, 145, 156, 158
and traditional owners, 48, 77, 164
urban spaces, 12, 48
worldview, 177
Indigenous knowledges
accumulated over millennia, 7, 19–20, 161; building, 30; of nutrition, 35; technology, 35, 79, 84, 85, 87–8
of Country, 13, 30–2, 46, 55, 60–1, 71, 74, 79, 98; educating others, collaboration, 46–7
elders, 10, 14, 30, 53, 74, 84, 89, 91, 157, 191 n 51; as libraries, 31
evidence, 21, 22, 23; gaps in, 96; and records of settlers/invaders, 38, 39, 71–2, 77, 82, 101, 124–5, 131, 133, 134, 137, 140, 142–3

ingenuity of, 44
integrated system, 2, 14, 31
knowledge holders, 10, 44, 45, 48, 62, 189 n 17; sharing, **32**, 91
and oral traditions, histories, 39, 77, 147; covering millennia, 39; transmission of knowledge, 41, 74
protecting, preserving, 1, 26, 74, 147–8, 161; and intellectual property, 148, 161
scientific knowledge, 14, 16, 19, 32, 37, 62, 160; ethnobotany, 27, 184 n 15; and method, 35, 41; paleoecology, 20, 64, 137
Indigenous science, 34, 35, 40, 160, 176; promoting knowledge, pedagogies, 43–4
relationship with non-Indigenous science, 41–2, 98; anthropology, 65–6, 93, 133, 138; archaeology, 22, 34, 38, 78, 82, 84, 96, 103, 120, 123, 125, 131, 135, 137, 139–42, 144–5, 153, 154, 156; and authenticity, 45, 66; and bias, 41, 46, 63, 64–5; collaboration, 147; knowledge undervalued, 35, 37, 39, 41, 44, 66–7; non-Indigenous ways privileged, 43; oppression of First Peoples and knowledges, 42–3, 44; valued by some, 34, 39, 43, 71, 82
and a voice, 63, 65, 66, 67
see also colonisation; plants
Irish, Paul, 103
Isaacs, Jennifer, 154

Jacob (Barkandjiman), **70**, 75, 90, 91, 99
Janke, Terri, 147–8

Jones, Jonathan, 13, 79–80
marrum (overflowing), 13
untitled *walam-wunga.galang (grindstones)*, 79
Jones, Rhys, 22

Kelly, Fiona, 72
Kennedy, Uncle Brendan, 30–1, 33
Kershaw, Peter, 22
Kimberley, East, 25–7, 28–9
Koorie Plants, Koorie People (Gott and Zola), 153–4

Lindsay, Robert, 72
Linnaeus, Carl, 107

Martin, Aunty Sarah, 96, 99, 191 n 51
Mary (Barkandjiwoman), **70**, 75, 76, 78, 81, 91, 99
Mathew, John, 102
Melbourne (Narrm), 51
  Birrarung (Yarra), 48, 51, 53, 56, **57**, 58
  Bolin Bolin, 51, **52**, 53, 54–6, **57**; degraded, 54; a jewel, 53; as meeting place, 52; pollution, 53; research, **57**, 137; rewatering, 53
  landscape since colonisation, 51, 53, **57**
  under Wurundjeri and Boon Wurrung/Bunurong: billabongs, 51–4, curated, managed landscape, 51, 53, 57–8, 59–60, 67–8; rainforest, 56, 57; species flourished, 51, 52; wetlands, grassy plains, 51, 59
Memmott, Paul, 147
Miller, Maddison, 33–4
Mitchell, Thomas, 77, 102

Morton, Oliver, 105
Moysey, Granny, 190 n 31
Muir, Pat, 28
Mulvaney, Ken, 139
Murray-Darling Basin, 154, 156
  and sustainability, 168–9
  and water scarcity, 110, 168–9
Murray-Darling Basin Royal Commission (SA), 10
Murray River, 100
  population density on floodplains, 102
museums, other institutions
  Indigenous artefacts displayed, 42–3, 133, 137, 140–1, 143, 144–5, 146
  and Indigenous peoples, 43, 146

O'Connor, Sue, 127, 136
Orwell, George, 65
Ouzman, Sven, 120

Page, Alison, 147
Pardoe, Colin, 153, 154, 157–9
Pascoe, Uncle Bruce, 6
  and Black Duck Foods, 171
  Dark Emu, 46, 79, 82
People of the Paroo River, The (Hope and Lindsay), 72
Péron, François, 106
Perth
  wetlands, 103
photography
  of Aboriginal people, 69, **70**, 71, 72, 73, 75
Pitman, Heidi, 131, 133, 134, 140, 141, 144–6
plants and Indigenous Australians
  as architects of, central to (Indigenous) life, 4, 16–17, 20, 29, 50, 71, 96–7, 105, 173

'bush food' and botanicals industry, 14, 160–1, 162, 163; Indigenous-led opportunities, 163–5; lack of benefit for First Peoples, 160–1, 162, 163, 166–7; naming, knowledge and property rights lost, 166–7, 172; non-Indigenous patents, 166, 167

and clothing, 29

connecting to Country, 1, 2, 13, 14; caring for, 50

cultural, spiritual dimensions of, 3, 120–1

define landscapes, 50, 130

eucalypts, 19; oil and Indigenous science, 40, 41

exploit animals and humans, 4, 17

and fire regimes, 25, 56–7, 58–9; ashes, 95

as food, 4, 29, 42, 76, 97, 123; breadmaking over millennia, 5, 79; nardoo, 35, 78; range of plants, 162, (many unrecognised), 168–9;and synergistic relationships: human, plant, animal, insect, 160

future foods, 168; and climate change, 168–9; dealing with invasive weeds, 169–70; indigenous plants, 169, 171–2

grasses, grasslands, 58–9, 78, 80, 169–70, 171; and grains, seeds, 59, 77, 80; pollen grains, 24; preparation, 77; support animals, 59

interactions with, knowledge, 1, 2, 6, 11, 13, 29, 35, 46, 48, 96, 120, 160, 167; and the Anthropocene, 110; and intellectual property, 6, 161, 165, 166, 167; and recognition, protection of knowledge, 161–3, 164, 165, 166, 167, 172; and sustainability, 6–7, 28, 161; and wider community,12, 48

invasive species, 29

invisibility of, 103, 114, 121, 130

lives of plants, 3; and mobility, 106

making use of plants, 69, **70,** 76; starch and glucose, 105; value of, 98

manipulation of: and fire, 20, 58–9; vegetation change, 22, 125–6

as medicine, 6, 29, 91–2, 97, 166

native plants and Australian conditions, 171; and nutrition, 171

and photosynthesis, 104–5

and reintroducing indigenous plants, 6; adapted to this climate, soil, 6; and food security, 6

revegetation, traditional food projects, 46

and sharing knowledge with all Australians, 173–4, 175; connection to Country, 174; and future generations, 174

and song cycles, 1

and stories, 14, 80, 103–4

and tools (stone): adzes, 81; axe heads, 81, 119; grindstones, **76,** 76–7, 79, 80, 119, 139, 140–2, 143, 156–7; and history of seed grinding, 137; irons, 81; rasps, 81; record of, 137, 138

traps and nets, 81–2, **82,** 83, 84–5, 90; fishing nets, **82,** 85

wheat, 17, 28

women's roles: in harvesting, preparation, 5, 24, 42, 80, 94, 95, 103–4, 142; chewing cumbungi, 84; erased from history, 5, 24, 103–4, 121

INDEX

wood/plants
  building materials for shelters, 2, 29
  canoes, 2, 97
  digging sticks (and women), 86, 90, 93, 94, 95, 96, 104, 143; and cultivation, 94–5; design, 94; spiritual power, 95
  fibres, 140, 144; for nets, bags, 2, 144
  and innovation, knowledge, 42–3
  resin, 1–2
  for tools, weapons, 1, 29, **70**, 76–7, 80, 97; boomerangs, 87–8; clubs, 86–7, 88; after colonisation, 86–7, 95; coolamons, **70**, 80–1; fire for strength, 87; and oils, fats, 87; spears, 89–90
  see also backyards; Country; cumbungi; spinifex; yams
Poelina, Anne, 43

quandong (genus *Santalum*)
  bitter quandong (*Santalum murrayanum*), 153–4
  collecting, 158
  conservation, protection, 159
  and cultural connection, 150; relationship between plants, people, animals, 156
  fruit of, 149, 150, 152; as cakes for storage, 158; drying, 158; harvesting, 158; nuts, 151, 153, 155, 156, 158
  impact of European settlement, 159
  importance of, 149, 159
  and knowledge, 151
  processed, 151, 152, 158, 159; stored, 151
  propagation, planting, 151
  range of uses, 151, 154, 156; medicinal, 152–3, 154, 158
  Santalum lanceolatum, 156
  and technology, 151, 156; nutcrackers, 157–8, (weight), 158
  trees, 149; hemiparasitic, 150, 159; landscape and distribution, 149, 154, 157–8, (clusters), 157, 158, (and emus), 155, 156, 157; prized for timber, 151
  see also plants

Rose, Deborah Bird, 65
Rots, Veerle, 137

Schutz, Hugo, 40–1
Simon, Biddy, 25, 26, 29, 121, 156
  and yams, 115, 116, 117, 118, 125
Smith, Mike, 36
spinifex (genus *Triodia*), 135, 138
  case studies: adhesives, hafting, 137–9; fibrecraft, 137, 139–44
  characteristics, 131
  collecting, processing: resin, 133–4; and exchange, 134; and women, 134, 142
  dominating grasslands, 130; species, 130, 131, 134, 135
  and grass tree (*Xanthorrhoea*), 135, 138
  and habitats: changes over time, 136
  Indigenous knowledge, 133, 134, 139, (and rights, recognition), 148; earliest evidence of use, 136; and technology, 135, 140
  management: fire, 131, 148
  new uses for: as biomaterial, 147; building, insulation, 147; and sustainable harvesting, 148

211

INDEX

range of uses for Aboriginal people, 132–3; ceremonial, 133; hafting, 132, 135; of leaves, fibres, 130, 131–3, 135; medicinal, 132, 133, 135; ornamental, 133; of resin, 130, 131, 133, 134, (cake of resin), 144–6; structural, 133
and settlers, 130, 131
and tangled lechenaultia (*Lechenaultiadivaricata*), 138
and Western science, 130, 131, 133, 134, 135, 136, 137, 139–42, 143; museum collections, 133, 137, 140–1, 143, 144–5
see also plants
Sturt, Charles, 77, 79–80
Sutton, Peter
Farmers or Hunter-Gatherers? (with Keryn Walshe), 67

Tasmania (lutruwita), 21
Aboriginal occupation, 23
Terry, Michael, 142, 148
Through a Land of Promise, 142–3

Veth, Peter, 120

Wallis, Lynley, 131, 133, 134, 140, 141
Wandanga, Polly, 25, 121
and yams, 115, 116, 117, 118, 119, 125
Wandin, Uncle Dave, 53
Williams, Nancy, 25

yams, 103
and climate change, 129
and Country, 114, 128, 129
fire management, 125–6; protecting yams, 126; rainforest, 126, 127
and humans: evolution, 120, 121; sustenance, 120
and landscape, 4, 27, 114, 115, 119, 122; disruption of access, forced removal of people, 125, 129; habitats, 115, 125, 128; impact of cattle, 125; storage, 128; transformation, 124–5, 127, 128; warran grounds, 124, 128; yamming and quarry sites, 119
quality, 118–19, 129
replanting, 118
species (Australian): *dangar* (*Dioscorea transversa*), 116–17, 123, 125; *garok* (*Dioscorea bulbifera*), 116, 117, 118, 123, 125, 127, 128, 129; Kulama, 122, 127; 'native yam' (*Dioscorea hastifolia*), 119–20, 123, 124, 125
and starch, 114, 120, 128
work of (mostly) women, 121–2; antiquity of, 120; with children, 121; collecting, 26–7, 114, 115, 116, 117, 118, 121, 122, 124, 127, 128, 129; digging process, 117, 118–19, 120, 121, 122, 123, 124, 128, (time consuming), 122, 123, 128; harvesting, digging rights, 125; and knowledge, 115, 121, 128, 129, (and transmission), 129
see also plants

Praise for the First Knowledges series …

'This beautiful, important series is a gift and a tool. Use it well.'
—Tara June Winch

'An in-depth understanding of Indigenous expertise and achievement across six fields of knowledge.'
—Quentin Bryce

'Australians are yearning for a different approach to land management. Let this series begin the discussion. Let us allow the discussion to develop and deepen.'
—Bruce Pascoe

'These books and this series are part of the process of informing that conversation through the rediscovery and telling of historic truths with contemporary application … In many ways, each individual book will be an act of intellectual reconciliation.'
—Lynette Russell

'The First Knowledges books are proving to be among the most fascinating and important that I have ever read; an astounding gift of wisdom delivered with generosity and optimism, offering no less than a new vision of what Australia is and what it can be … They deserve to change minds, lives, and hopefully the development of Australia itself.'
—Jez Ford

The best of both worlds

TITLES IN THE FIRST KNOWLEDGES SERIES

SONGLINES
Margo Neale & Lynne Kelly
(2020)

DESIGN
Alison Page & Paul Memmott
(2021)

COUNTRY
Bill Gammage & Bruce Pascoe
(2021)

ASTRONOMY
Karlie Noon & Krystal De Napoli
(2022)

PLANTS
Zena Cumpston, Michael-Shawn Fletcher & Lesley Head
(2022)

LAW
Marcia Langton & Aaron Corn
(2023)

Published in conjunction with the National Museum of Australia
and supported by the Australia Council for the Arts.